NASA MISSION AS-508

APOLLO 13

1970 (including Saturn V, CM-109, SM-109, LM-7)

Dedication

To the four flight-control teams who nursed Apollo 13 through almost six days of an unforgettable mission, where the finest principles of stoic determination and human resolve saved the lives of three very brave men.

First published in October 2013

A catalogue record for this book is available from the British Library.

ISBN 978 0 85733 387 2

Library of Congress control no. 2013941310

Published by Haynes Publishing,
Sparkford, Yeovil,
Somerset BA22 7JJ, UK.
Tel: 01963 442030 Fax: 01963 440001
Int. tel: +44 1963 442030
Int. fax: +44 1963 440001
E-mail: sales@haynes.co.uk
Website: www.haynes.co.uk

Haynes North America Inc.,
861 Lawrence Drive, Newbury Park,
California 91320, USA.

Printed in the USA by Odcombe Press LP,
1299 Bridgestone Parkway, La Vergne,
TN 37086.

Cover cutaway by Ian Moores

How to use this book

During the Apollo programme, NASA measured distances in nautical miles, one nautical mile being 6,076ft (1,852m) which is one degree of latitude at the meridian. However, one statute mile consists of 5,280ft (1,609m), a unit of measurement set by Parliament in 1593, and this is the one with which most readers preferring Imperial measurements are more familiar. Accordingly, all the distances used in NASA documents and flight records using nautical miles are different from those in this book using statute miles. To convert to nautical miles, multiply the statute miles in this book by 0.8689.

Also, the NASA convention of recording the time of events in hours, minutes and seconds from lift-off is adhered to, this being ground elapsed time (GET), the zero time at the launch pad being expressed as 00:00:00 from which flows the resulting time in this book. At frequent intervals, however, the more conventional 24-hour clock is used for events in local Houston time to give readers a frame of reference as to the time of day in Mission Control.

About the author

After obtaining his PhD in Earth and Planetary Physics from a scholarship programme run by Sen Clinton P. Anderson, Dr David Baker worked with NASA between 1965 and 1990. He was involved with Apollo 13 in Mission Control during the six-day flight. He was witness to events in the Mission Evaluation Room and helped carry out verification checks on some of the consumables calculations vital for returning the crew safely to Earth.

David also participated in activities across the mission-support rooms. After the flight he was asked to prepare analytical evaluations for Apollo contractors. Subsequently, and as a result of his experiences in Mission Control during the near-fatal flight, he prepared analyses of mission-failure mode mitigation work that led to new analytical tools for risk analysis.

He has written more than 90 books on space science, engineering and technology, and on the politics of the space programme, serving seven years as the editor of *Jane's Space Directory*. In 1987 he was elected a member of the prestigious International Academy of Astronautics and is a Fellow of the British Interplanetary Society.

David is the recipient of the 1998 Rolls-Royce Award for Aerospace Journalist of the Year and the 2008 Sir Arthur Clarke Award. He has made numerous TV appearances and is regularly interviewed by media stations around the world. He is currently the editor of *Spaceflight*, the monthly news magazine of the British Interplanetary Society.

NASA MISSION AS-508
APOLLO 13

1970 (including Saturn V, CM-109, SM-109, LM-7)

Owners' Workshop Manual

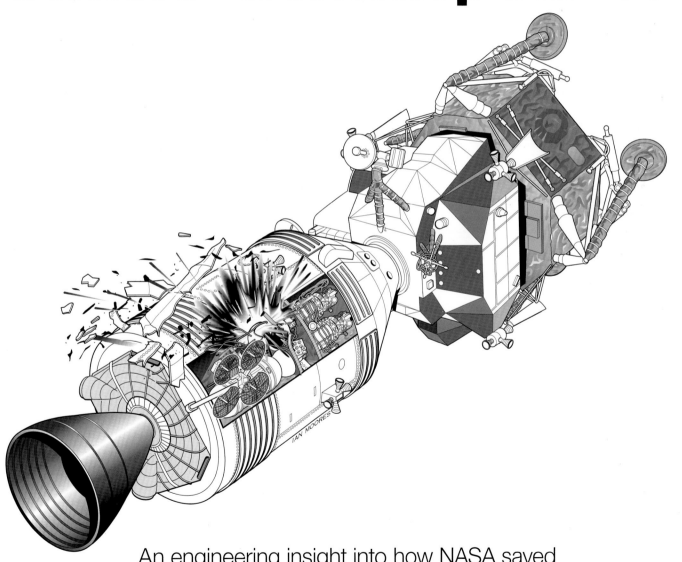

An engineering insight into how NASA saved
the crew of the crippled Moon mission

David Baker

Contents

OPPOSITE Apollo 13 stands ready for flight at Launch Complex 39A for NASA's third landing in nine months, to a place on the Moon of great scientific interest. *(NASA)*

Chapter One

Prelude

◀━━━(●)━━━━━━━━━━━━━━━━━▶

Confidence was high. After two missions to the Moon to prove the landing systems worked, Apollo 13 was to be the first mission to conduct real science and examine a region of the Moon difficult to get to. A place that could not have been reached without the two previous flights raising the level of assurance that precision guidance across difficult surface features would get the Lunar Module down safely.

OPPOSITE Ten days before Christmas Day, 1969, Apollo 13 was moved from the Vehicle Assembly Building at NASA's Kennedy Space Center to Launch Complex 39A, from where it would be launched to the Moon almost three months later. Originally scheduled to fly in late 1969, when Apollo 11 successfully landed on the Moon the pace was slowed and launch was set for 12 March 1970 before being set back to 11 April. *(NASA-KSC)*

LEFT The mission badge for Apollo 13 was designed by Lumen Martin Winter, a famous painter noted not only for his murals but for sculpture. Based on conversations with Lovell, Mattingly and Haise, he depicted three horses pulling a sun chariot from the Earth to the Moon, a circular disc surmounted with the words 'Apollo XIII' and the words in Latin 'Ex luna, scientia' – 'from the Moon, knowledge'. An appropriate sentiment for this mission, the first to fully turn from engineering demonstration flights to science and the exploration of lunar geology. *(NASA)*

BELOW Officially named on 6 August 1969, the prime flight crew consisted of James B. Lovell, Commander (CDR), Thomas K. Mattingly, Command Module Pilot (CMP) and Fred Haise, Lunar Module Pilot (LMP). The back-up crew comprised John Young, Jack Swigert and Charles Duke, the support crew being Jack Lousma, Vance Brand and William Pogue. *(NASA)*

Planning for Apollo 13 as a potential mission to Fra Mauro began with the successful landing on the Moon by Apollo 11 in July 1969. Originally, the Apollo Command and Service Module had been named after the constellation *Aurigae*, but that had a phonetic similarity to *Aquarius*, the name of the Lunar Module, so that call-name was changed to *Odyssey*. The mission badge for Apollo 13 had been approved early in the year, and publicly announced by NASA on 23 January.

Training for Apollo 13 began in January 1969 when Jim Lovell, William A. Anders and Fred Haise were selected as back-up crew to Apollo 11. In May 1969 Anders was nominated by President Nixon to head the National Aeronautics and Space Council but he served out his assignment as Apollo 11 back-up Command Module Pilot before leaving NASA. Apollo 13's crew would get Mattingly in his place for the third lunar landing attempt.

Changes to Apollo crews were not uncommon and could be triggered by the most unpredictable events. In 1964 America's first astronaut in space Alan Shepard was removed from flight status when he was diagnosed with Ménière's disease, an affliction to the inner ear that causes nausea and disorientation. Biding his time as chief of the astronaut office, Shepard responded successfully to surgical treatment using a newly discovered technique, and in May 1969 he was restored to flight status.

Keen to get back in space, Shepard was initially considered for assignment to Apollo 13 along with Stu Roosa and Ed Mitchell, but that crew was put back to command of Apollo 14 to allow more time for training. But for that they would have been the Apollo 13 crew.

It had taken more than two years to manufacture, but the Apollo spacecraft (CSM-109) had been at Kennedy Space Center (KSC) since 26 June 1969 and the Lunar Module (LM-7) arrived two days later. With the pace of assembly and launch preparation slowing noticeably in the wake of Apollo 11's successful Moon landing, altitude tests on CSM-109 were completed by mid-September. A similar evaluation on *Aquarius* cleared

the LM a week later, and on 9 December the spacecraft were moved to the Vehicle Assembly Building (VAB) and stacked atop the Saturn V the following day.

BELOW James (Jim) Lovell was born on 2 March 1928, and following a distinguished career as a naval aviator he joined NASA as an astronaut in September 1962. Captain Lovell was selected as co-pilot to Frank Borman on the 14-day Gemini 7 flight in December 1965. In this official portrait taken in 1966 he is seen with a model of the Gemini-Titan II stack. Lovell was chosen to command the two-man Gemini 12 mission with co-pilot Edwin 'Buzz' Aldrin flown in November 1966, and served as Command Module Pilot on the epic journey of Apollo 8 in December 1968, the first flight to leave Earth's gravity and the first to orbit the Moon. Lovell retired from NASA in 1973 and has since carved out a highly successful business career. *(NASA)*

ABOVE Thomas K. (Ken) Mattingly II was born on 17 March 1936 and served as an aviator with the US Navy from 1958 until selected as an astronaut in April 1966. Captain Mattingly was selected for Apollo 13, but never flew the mission due to exposure to rubella, serving instead as support for the rescue mission back from around the Moon. Mattingly was selected as Command Module Pilot for the Apollo 16 mission, flown in April 1972 as the penultimate lunar-landing flight, and went on to conduct two Shuttle flights, commanding the second mission, STS-2, in June 1984 and the STS 5I-C flight in January 1985. He retired from NASA that year and left the Navy in 1986 with the rank of Rear Admiral, entering the employ of a national security advisory and analysis company. (NASA)

ABOVE Fred Wallace Haise Jr was born on 14 November 1933 and began a career in naval aviation in 1952 prior to joining NASA as a research pilot at the Lewis Research Center in Cleveland, Ohio, in 1959. From 1963 he served as a research pilot at NASA's Flight Research Center at Edwards, California, and in April 1966 Haise was selected as an astronaut, serving as back-up Lunar Module Pilot for Apollo 8 and Apollo 11. From 1973 to 1976, Haise was technical assistant to the manager of the Space Shuttle Orbiter, and commanded two Air Launched Test flights in 1977, flying the Orbiter Enterprise off the top of a converted Boeing 747 to demonstrate its handling qualities. Fred left NASA in June 1979 and joined the Grumman Aerospace Corporation, builders of the Lunar Module, as a test pilot, retiring in 1996. (NASA)

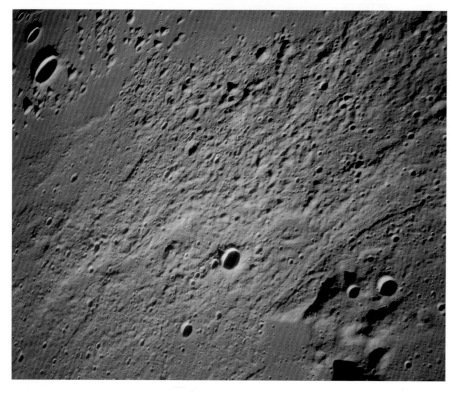

LEFT The site selected for Apollo 13 was known as Fra Mauro, situated at 3º 40' 7" S by 17º 27' 3" W, located 111 miles (179km) east and slightly south of the Apollo 12 landing site. The prime meridian in lunar cartography lies upon a north-south line of longitude bisecting the Moon as viewed from Earth. The site was named after the crater Fra Mauro, approximately 60 miles (96.5km) in diameter, 30 miles (48.3km) to the south of the spot selected for touchdown. It was in a hilly upland region, the entire area known as the Fra Mauro formation comprising a massive ejecta blanket of material thrown out from the formation of Mare Imbrium, a giant basin 720 miles (1,158km) in diameter which lies far to the north. (NASA)

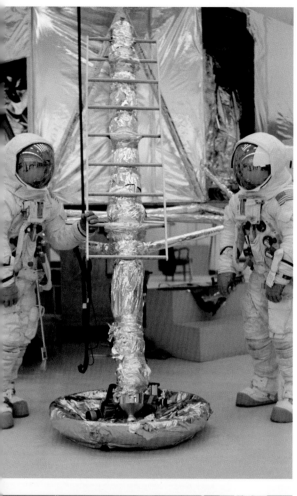

FAR LEFT Jim Lovell (left) and Fred Haise practise suited operations around a 1g mock-up of the Lunar Module at the Kennedy Space Center in early 1970. *(NASA)*

LEFT Equipped with a 16mm camera, Fred Haise rehearses activity with a tool carrier, for a geological exploration he planned to conduct on the Moon during one of two space walks it would be his fate never to perform. *(NASA-KSC)*

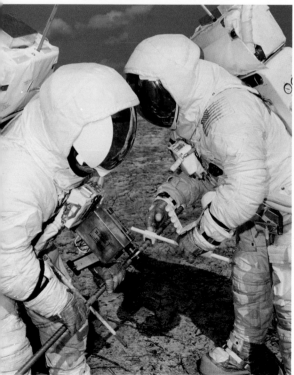

LEFT Lovell and Haise practise working with core tubes carried aboard Apollo 13 on what was the first truly geological expedition, following two essentially engineering flights in which the capabilities of the Apollo system were put to the test. *(NASA-KSC)*

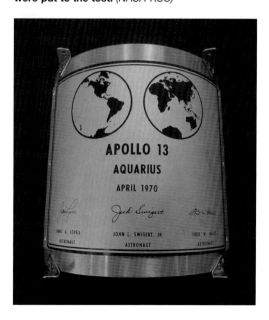

LEFT Cheated of an opportunity to fly on Apollo 13, Mattingly's place aboard *Odyssey* was taken by Jack Swigert, whose name is inscribed on the plaque that would have adorned the forward landing leg of Lunar Module *Aquarius* on the surface of the Moon. *(NASA)*

Chapter Two

To the Moon!

Last-minute problems beset this third planned Moon landing. A crew member stood down for fear he might have contracted German measles, trouble hampered pre-flight preparations for critical systems on board *Odyssey* and then the Saturn V rocket had technical difficulties that threatened an abort on the way to orbit. But despite these potential show-stoppers, Apollo 13 finally got on its way, leaving Earth for the Moon as planned.

OPPOSITE A view looking forward as *Odyssey* turns around and faces the top of the S-IVB stage, with the lower segment of the SLA and *Aquarius* nestled inside. *(NASA)*

Apollo 13 was moved from the Vehicle Assembly Building to Launch Complex 39A on 15 December 1969, a month before the decision to set back the launch from 12 March to 11 April 1970. In preparing for flight, Apollo-Saturn ground personnel would go through every step in the sequence up to but not including the point of ignition, a procedure known as the Count-Down Demonstration Test (CDDT). This is divided into a wet CDDT, during which the Saturn V and Apollo spacecraft is loaded with propellants but without the crew present and only essential personnel in attendance; and a dry CDDT, without propellants but with the crew aboard.

BELOW Saturn V AS-508 carried *Aquarius* (LM-7) encapsulated inside the four panels of the Spacecraft Lunar Module Adapter (SLA) supported on the lower segment of the SLA, on top of the third (S-IVB) stage of the rocket. *Odyssey* (CSM-109) was positioned above the LM, on top of the SLA. *(North American Rockwell)*

ABOVE On the late afternoon of 10 April, 1970, the giant Mobile Service Structure (MSS) was pulled away from Launch Complex 39A, clearing the pad where Apollo 13 awaited lift-off. This view was taken from one of the gantry enclosures on the MSS as it withdrew. *(NASA-KSC)*

RIGHT The Saturn V (AS-508) on the morning of launch. *(NASA-KSC)*

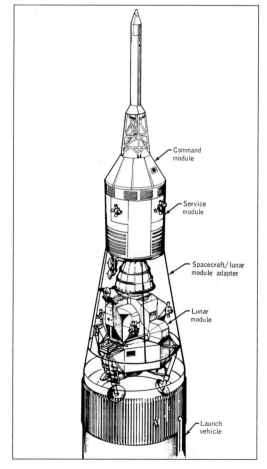

Command module

Service module

Spacecraft/lunar module adapter

Lunar module

Launch vehicle

On 25 March, the wet CDDT was completed with the offloading of cryogenic fluids. At atmospheric temperature, the RP-1 (kerosene) would remain in first stage of the Saturn V (the S-IC stage). But as standard for these procedures, ground technicians emptied 39,000 litres of oxygen – used for pre-cooling the rocket's plumbing – into a ditch adjacent to launch pad LC-39A. Pushing the normal atmospheric mixture of 78% nitrogen and 21% oxygen aside, on this occasion the super-rich oxygen gas began igniting the hot engines of police and security cars nearby, and temporarily confusion reigned!

That was not all. During the CDDT, cryogenic oxygen was also loaded aboard the two oxygen tanks in the Service Module. These tanks provided oxygen to pressurise the Command Module, with about two-thirds of the total quantity allocated to the three fuel cells which, when mixed with hydrogen from two separate tanks, produce electrical power and water as a by-product. Each tank contained a temperature sensor with a small heater and two fans. When the time came to empty both tanks to 50%, tank No 2 would not drain below 92%; only tank No 1 bled down through the vent valve.

ABOVE LEFT When Ken Mattingly was exposed to rubella, back-up Command Module Pilot Jack Swigert (shown) was assigned his seat in *Odyssey*. Born on 30 August, 1931, John L. (Jack) Swigert Jr served in the Air Force as a fighter pilot, and then as a test pilot with North American Aviation before joining NASA as an astronaut on the third try in April 1966. He was appointed a member of the Apollo 7 support crew and was assigned as back-up Command Module Pilot for Apollo 13 before flying this mission in place of Ken Mattingly. After Apollo 13, Swigert took leave to become executive director of the House Committee on Science and Technology, resigning from NASA in 1977. He was elected to the House of Representatives but died of cancer on 27 December 1982, eight days before taking up office. *(NASA)*

ABOVE Although the prospect of Swigert joining the flight crew for Apollo 13 came up a mere five days before launch, NASA lost no time in having the new crew pose for publicity shots. Here, the life-long bachelor Jack Swigert (left) holds a model of *Odyssey*, while Jim Lovell holds *Aquarius* and Fred Haise tries to look interested. *(NASA)*

The problem was thought to be minor and the problem was set aside.

The following day, 26 March, the dry CDDT went off without a hitch and was successfully completed with only minor discrepancies. Further attempts to drain down the No 2 tank were made next day, and by trying to offload it through the fill line technicians managed to get it down to 65%. In the belief that oxygen was leaking back up the fill line from the vent tube, a standard de-tanking procedure was started. Nothing happened, leaving the only solution being to boil off the oxygen by switching on the internal heaters, warming the cryogenic liquid to a gas and then turning the fans on. This procedure continued on 28 March.

For more than 90 minutes, KSC technicians out on the Launch Umbilical Tower at LC-39A, high above the concrete pad, fed power to the heaters from the ground support equipment, using a 65V DC power source. When after some 4½ hours the quantity still read 35% it was decided to start a pressure cycle, but it took five such cycles to empty the tank after eight hours' continuous operation. With such erratic performance from the tank, technicians had to decide whether it was up to the job.

Two days later, on 30 March, they decided to perform an aggressive technique, filling it with liquid oxygen and then forcing in gaseous oxygen to the 20% level, converting the gas to a partial liquid. Again, it could only be emptied

through a repeated series of pressure cycles with the heaters on over several hours, once again using the 65V DC power of KSC ground equipment.

Oxygen was critical to maintaining life aboard the Command Module and as a reactant for the fuel cells to produce electrical power essential to running the spacecraft and keeping the crew alive. Engineers were consulted and the decision was made. The tank probably had a loose or ill-fitting fill tube, they said. Had it not had an incident some time before? The history of this tank had been at best a chequered one, a story none of the technicians knew as they wrestled with the problems on LC-39A.

A flawed oxygen tank

The story of tank No 2 really began on 26 February 1966 when North American Aviation contracted with Beech Aircraft Corporation to design, develop and build cryogenic tanks for the Service Modules. Oxygen tank No 2 which was ultimately installed

Blowout disc
Closeout cap
Supply line
Temperature sensor
Fan motor
Thermostat
Heater
Insulation
To fuel cell/ECS
Capacitance gage
Fan motor
Pressure transducer
Pressure switch
Relief valve
Closeout cap
Overboard

LEFT During preparations for the Apollo 13 mission, serious problems arose with the No 2 oxygen tank installed in sector 4 of _Odyssey_'s Service Module. Oxygen tanks carried two small fans driven at 1,800rpm for stirring up the cryogenic liquid, and a heater for controlling the balance between slush and liquid. The capacitance gauge provides a measure of the quantity remaining. Each oxygen tank has a nominal quantity of 320lb of liquid oxygen at a pressure of 865–935lb/in² (5,964-6,447kPa) filled at -297°F (-182.8°C). The tank operates at a temperature range of –340°F (–206.7°C) to +80°F (26.7°C). A pressure-relief valve operates at a nominal 1,000lb/in². The blowout disc was designed to fail at 2,200lb/in² (15,169kPa) at –150°F (–101.1°C). The disc was open to the vacuum between the inner and outer pressure vessels, and was designed to vent an over-pressurising tank before the outer shell split and caused collateral damage. Under extreme conditions, where the temperature increased rapidly, it could be overwhelmed, and the tank itself would fail. (Beech Aircraft)

RIGHT The design of the tank was sensitive to damage caused by incorrect fitting, rough handling and electrical failures resulting in short circuits. Wiring for the capacitance gauge and temperature sensor and associated fan motors is attached via the exterior of the quantity probe to a conduit in the dome. The abundance of electrical wiring in the highly pressurised container left little room for failure containment, a catastrophe almost inevitable from an otherwise inconsequential failure of some minor component. *(Beech Aircraft)*

BELOW The two oxygen tanks were essential for maintaining a life-sustaining atmosphere in the Command Module and for providing a reactant gas to the three fuel cells (see Chapter 3). The tanks themselves were on the critical path for controlling the environment and providing electrical power aboard the spacecraft. This diagram explains how the ground-support equipment (GSE) provided a fill-and-replenishment capability, and how the supply lines delivered oxygen to the fuel cells. *(North American Rockwell)*

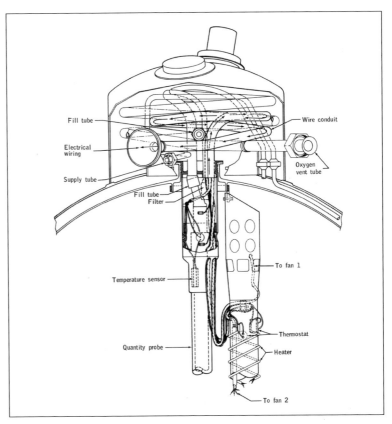

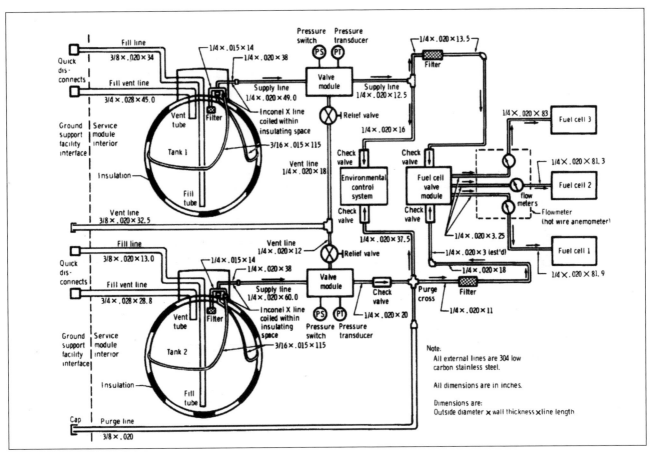

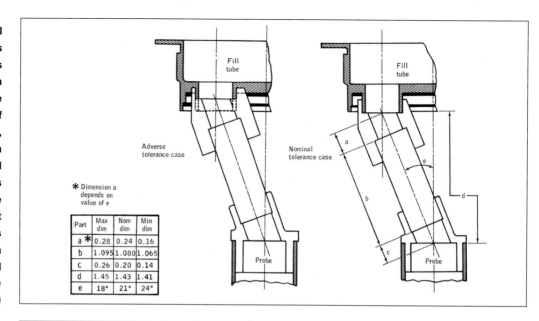

✱ Dimension a depends on value of e

Part	Max dim	Nom dim	Min dim
a ✱	0.28	0.24	0.16
b	1.095	1.080	1.065
c	0.26	0.20	0.14
d	1.45	1.43	1.41
e	18°	21°	24°

RIGHT On 21 October 1968, the oxygen tanks installed in SM-106 were removed for modifications. After lines and wires had been disconnected, the bolts holding the shelf in place were removed and a fixture suspended from a crane was employed to extract it from sector 4. However, one bolt was mistakenly left in place, and this caused the shelf to rise approximately 2in (5cm), breaking the fixture and allowing it to fall back. It appears that upward forces may have caused the dome of the tank to strike the underside of the upper shelf. *(North American Rockwell)*

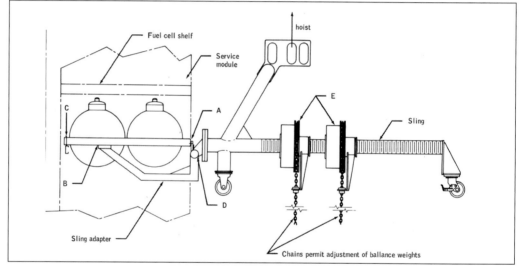

Chains permit adjustment of ballance weights

in SM-109 bore serial number 10024XTA0008. In post-manufacturing tests numerous flaws were discovered in what eventually became tank No 2 in SM-109, and some re-welding was required when it was found that incorrect wire had been used. Then a noisy electric fan motor that drew excessive power got attention and it was taken apart and reassembled with a new motor, fan and heater.

Acceptance tests at Beech included leak tests at 500lb/in² (3,347kPa) and proof testing at 1,335lb/in² (9,205kPa) using the tank heaters to 'warm' the liquid oxygen and raise the internal pressure using a standard 65V DC power supply. However, while it was possible to de-tank the liquid oxygen the rate of heat leak worried engineers. It was too high, but, after

some rework, although still out of spec, it was accepted with a formal waiver. But there were other discrepancies too. Electrical plugs were installed in oversize holes, oversize rivets were noted in the heater assembly, and there was concern about the integrity of the seal between the two tank shells.

Nevertheless, on 3 May 1967 it was shipped off to North American after being filled with helium at 5lb/in² (34.4kPa), a standard procedure for transporting cryogenic tanks. At NA it was paired with tank number 10024XTA0009, which was designated tank No 1, and attached to shelf 0632AAG3277, where it was assigned to Service Module 106 for the Apollo 10 flight as oxygen tank No 2. That installation took place on 4 June 1968 and tests

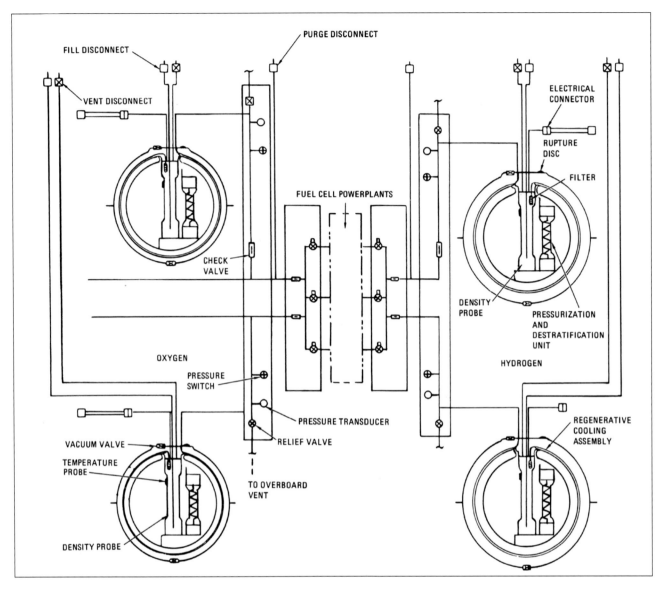

ABOVE The oxygen and hydrogen sides of the reactant supply to the fuel cells had parallel feed and line connections. *(North American Rockwell)*

showed no abnormalities, apart from a faulty valve which had to be replaced earlier.

Because engineers had noted an electromagnetic interference problem with vacuum pumps on earlier tank domes, modifications had been made. The decision to introduce those modifications to existing tanks necessitated removal of the shelf, and that was completed on 21 October. During the process of extracting the tank from the Service Module a bolt was inadvertently left in place, causing the shelf to rise 2in (5cm), the dome of tank No 2 striking the other shelf, before falling back.

The incident was recorded and tests conducted to see if the forces involved exceeded design limits, which they did not. But nobody could know for certain whether internal damage had been incurred. There was no way of looking inside to see if the precise arrangement of fill tubes, fans, heaters and drain tubes had been displaced by the shock. Nevertheless, the tank was deemed useable, and after the modification on 22 November 1968 it was installed in SM-109 for Apollo 13.

Technical glitches

None of this troublesome history was known to the technicians out on the pad in March 1970. With only five days to go before the countdown was due to start, pressure was mounting for a decision. It would take at least 45 hours to remove and replace the oxygen tank shelf in the Service Module, with the potential for incurring damage to adjacent systems. Late in the day on 30 March, 12 days before launch, it was decided to conduct a sequence of tests with the two oxygen tanks, running operations in parallel for comparison.

Again, there was no problem cycling each tank with gaseous oxygen, and tank No 1 emptied as designed. But when it came to emptying tank 2, once again it could only be completed by repeated pressure cycling with the heaters turned on. The problem was of serious concern, because the system itself was on the critical path for life support. Engineering personnel were brought in, together with managers at KSC, the Manned Spacecraft Center, from North American, Beech and NASA headquarters.

Key factors that would have triggered alarm bells and actioned a rapid response to replace the oxygen tanks, complete with shelf, were unknown to those responsible for making the decisions. Some people did not know about the prolonged application of power from the ground support equipment, others were unaware of the tank's manufacturing problems, a few had not seen the report on the shelf incident, jarring and possibly displacing internal components. Weighing all known factors, it was decided to proceed toward launch. Apollo 13 would fly with an erratic and temperamental oxygen tank.

But oxygen tank No 2 was not the only piece of hardware troubling technicians. Another tank containing supercritical helium (SHe) for pressurising the Lunar Module Descent Engine (LMDE) on *Aquarius*, was also giving trouble. The boiling point of helium is almost absolute zero (-452.2°F, or -269°C), due to weakness in the attraction between atoms because it is a noble gas. It was because it will not chemically bond with the hypergolic propellants for the Lunar Module Descent Engine that the engineering challenges in carrying this super-cold liquid were both necessary and unavoidable.

The SHe tank was spherical in shape with a diameter of 32.9in (83.6cm). It had a volume of 5.9ft³ (0.17m³) with initial liquid helium fill weight of 48.5lb (22kg). In preparing for any flight, the first time the tank is pressurised after installation in the Lunar Module is at a simulated time of T-48 hours during the CDDT. It is allowed to cold soak for approximately 28 hours, after which the tank is topped off. Natural heat leakage from outside the cryogenic tank gradually warms the helium, which increases internal pressure at an initial rate of 9.2lb/hr (4.17kg/hr). This usually falls to 7.5–8lb/hr (3.4–3.6kg/hr) by T-20.5 hours when topping off operations are completed.

During the cold soak at the CDDT for Apollo 13 the SHe tank showed an initial leak rate of 17lb/hr (7.71kg/hr), dropping to a fraction under 10lb/hr (4.54kg/hr), which is the specified limit. It was surmised that gaseous nitrogen, which had been kept in the SHe tanks to inhibit corrosion, had leached into the vacuum between the inner and outer pressure vessels during the five years since manufacture. It was this, they thought, that was acting to transfer heat into the tank at a higher than expected rate.

The supercritical helium for pressurising the descent engine propellant tanks is used only after ambient helium from a smaller tank – 14.9in (37.8cm) in diameter, containing 1.1lb (0.5kg) of helium – has been used to start the engine. The first time it was expected to be used on Apollo 13 was at an elapsed mission time of about 103hr 30min, when the engine was ignited to take *Aquarius* down to the surface of the Moon. Normal maximum supercritical helium storage pressure was 1,555lb/in² (10,722kPa), with a maximum of 1,710lb/in² (11,790kPa) at a density of 8.2lb/ft³ (131kg/m³).

A burst disc was built into the tank so that overpressure would be released via a non-propulsive vent. The vent comprised two diametrically opposed, oval-shaped holes. This was to ensure that the escaping gases would cancel each other out, so that the LM would not be thrown into gyrations should the burst disc rupture unexpectedly. The burst disc was designed to rupture at a pressure of 1,967lb/in² (13,562kPa).

At launch the SHe tank would be pressurised to just under 400lb/in² (2,758kPa) and, at a pressure rise rate of 10lb/hr (4.54kg/hr), it would reach approximately 1,430lb/in² (9,860kPa) by the time it was used during the descent phase at 103.5 hours elapsed time. It was deemed not to be a threat to the flight because *Aquarius'* descent engine was expected to be used for landing on the Moon long before the pressure reached a dangerous level, exhausting the helium from the tank as it forced propellants to the flow control valves.

Breathless preparation

The formal six-day countdown for Apollo 13 began on the morning of Sunday 5 April, the day astronaut Charlie Duke was diagnosed with rubella – German measles. Duke had been in contact with several astronauts, including the Apollo 13 crew, during the incubation period. NASA's chief physician Dr Charles Berry ordered tests on Lovell, Mattingly and Haise to see if they were immune. Although Jim Lovell's son had gone down with rubella only Mattingly was dangerously exposed, the others having had the disease already. But nobody would know whether Mattingly had contracted rubella until after launch.

The following day, Dr Berry recommended removing Mattingly from the flight list. Duke was back-up to Fred Haise, so a straightforward crew switch was impossible. With five days to launch, what should they do? The only possible option was to switch Mattingly for Jack Swigert, the bac-kup Command Module Pilot (CMP), who was also immune. Nobody liked breaking up a crew, and to put in a replacement just five days before launch was asking for trouble.

Ultimately, it was Jim Lovell's call. Only he could make the final decision as to whether Swigert was sufficiently up to speed as back-up CMP to take his place on the centre couch. In truth, if anyone had to go down with rubella it was better for the CMP to do so than either the Commander (CDR) or the Lunar Module Pilot (LMP). The CDR and the LMP would work together piloting *Aquarius* down to the surface and conduct two periods outside *Aquarius* exploring the surface. The CDR was the only crew member who would actually 'fly' both the Apollo spacecraft and the Lunar Module. The CMP was 'co-pilot' to the Commander in *Odyssey* and the LMP was 'flight engineer' to the CDR in *Aquarius*.

But it really came down to whether Lovell was happy with Swigert's performance as a member of his crew, and only after a thorough workout in the Command Module simulator would that be known. Spacecraft simulators are useful for familiarising the crew with spacecraft systems and with gaining experience in working closely together in a full-scale replica of their vehicle. But their real value was in fleshing out how to counter problems thrown at them by engineers controlling the simulator – a particularly cantankerous and appeal-resistant group with attitude, known as simulator supervisors or sim-sups (pronounced 'simsoops').

Theirs was the job few would want or be capable of carrying out, a task devoid of compassion, throwing every conceivable problem at the crew, usually in multiples of potentially catastrophic events, and recording every minute move made by the astronauts to get out of trouble. It required a certain type of rascally-minded individual with an intuitive engineering brain for fleshing out human weaknesses or faults, and then seeing how they extricated themselves!

This was what Jack Swigert survived when he, along with Lovell and Haise, spent several hours in the Command Module simulator at the Kennedy Space Center. With Flight Crew Operations Branch chief Riley McCafferty throwing everything at them, the crew was given a rough shakedown ride during which they were thrown every contingency, emergency and failure it was possible to dream up. When they emerged, McCafferty went into a huddle with Donald K. 'Deke' Slayton, director of flight crew operations, and declared that if Jim and Fred were happy, Jack was good to go. They were, and he was. It had been a breathless preparation that delivered a competent replacement crew member.

Before senior managers could sign off Jack Swigert into the centre seat of *Odyssey*, there would be a couple more days of simulator runs and technical analysis of how the crew responded, integrated and worked through problems together. NASA was circumspect and waited until 36 hours before launch to give formal approval. The day before the flight, the world got to know. It did little to restore interest in this third Moon landing, at least so far as the media was concerned.

The last-minute switch in crew seats prompted negative comment to the effect that 'It is an untimely reminder', said the *Philadelphia Inquirer*, 'of how much America spends to send men to the Moon and how little to combat disease on Earth.' Even the *Wall Street Journal* summed up the attitude of the nation in believing that few people 'really seem to care', and that the mission was generating 'little excitement at KSC or elsewhere across the land'.

That was not entirely the case. In all, around 700 journalists were at the Cape, albeit only a quarter as many as had flocked to see the launch of Apollo 11 just nine months before. But little boats and small watercraft filled the Banana River and the Indian River and 500 light aircraft flew in to Brevard County's private airstrips, while 25,000 cars and campervans gave police a hard time unobtrusively corralling spectators.

In nearby Cape Canaveral, and even down at Cocoa beach, billboards and signs were festooned with 'Good Luck!' messages. But this was the Space Coast, where jobs were generated by NASA and by the Air Force, where families were immersed in space talk and rockets and where livelihoods depended on a robust space programme. Around 100,000 people arrived to see the Saturn fly – perhaps a tenth of the numbers who crammed every spare standing spot around the Kennedy Space Center for the first Moon landing. But NASA had logged only 350 newsmen to cover the flight, of which 55 came from overseas, while Houston TV failed to carry the launch live.

RIGHT Jim Lovell gets suited up on the morning of launch, prior to a pressure -integrity check. *(NASA-KSC)*

Launch time

A phenomenal 4,500 people watched the launch of Apollo 13 from the VIP stands, with a further 7,000 out at a guest site on the causeway, the largest number at that location for any manned launch. As 11 April dawned weather reports for the day were good. Out on the launch pad a gentle breeze was washing in from the Atlantic and blue skies were tinged with only trace skeins of cloud. Temperatures would nudge up to 75°F and there was every prospect of a good flight. Set by the relative geometry of the Earth and Moon, the launch was set for a comfortable 2:13pm EST.

All through the night preparations had been under way, bathing LC-39A in a blanket of light. The count picked up at 4:13am and cryogenic oxygen and hydrogen began to flow into the tanks of the giant Saturn V. The RP-1 kerosene had been in the S-IC fuel tank since before the CDDT.

In the Mission Operations Control Room (MOCR) at Mission Control Houston (MCC-H), the first team of flight controllers were at their consoles at 2:00am local time (3:00am at the Cape) for the start of pre-launch procedures. Houston would be responsible for the vehicle after it ascended beyond the launch tower. Until then, KSC was in charge. The four MOCR teams had colours to denote their shift. The Black Team under flight director Glynn Lunney, a veteran flight controller and one of the founding fathers of the Manned Spacecraft Center, would be on duty until T-2 hours.

The crew spent the night before launch in the Manned Spacecraft Operations Building (MSOB) in the administrative area of the Kennedy Space Center. Woken just before 9:00am by Tom Stafford, chief of the Astronaut Office, they had a brief medical examination where they were passed fit for flight. Then it was time for a traditional breakfast of tenderloin steak, eggs, orange juice, coffee, and jelly and toast, followed by a short briefing on the countdown and final procedures. Shortly thereafter the crew suited up and left the MSOB at 11:07am, travelling eight miles (13km) to the launch pad in the familiar white transfer van.

Taking the elevator up to Level 9, they walked across the gantry 30 storeys above

the ground and entered the White Room, a box-like enclosure enveloping the Command Module. Jim Lovell was first to enter *Odyssey* at 11:32am, feet first on to the centre couch before sliding across to the left couch, where the closeout team led by Günter Wendt strapped him in. Haise followed next at 11:38am, first on to the centre couch then across to his position on the right-hand side of the spacecraft, before Swigert entered straight on to the centre couch at 11:44am, T-2hr 29min.

The only problem was difficulty in removing their protective helmet covers after they were strapped in, the limited flexibility of their arms in bulky space suits making this awkward. Meanwhile in Houston, the Maroon Team under Apollo 13 lead flight director Milton Windler replaced Lunney's people at the consoles.

With the spacecraft hatch sealed, at T-90 minutes the closeout crew fitted a segment of the Boost Protective Cover (BPC) that would shield the windows and the exterior of the Command Module from the blast of the Launch Escape System (LES) should it be used to carry the spacecraft to safety in the event of a booster malfunction. Test supervisor Clarence 'Skip' Chauvin carried out a series of cockpit checks with the crew before the final closeout

was completed at T-60 minutes. Only a few minor problems had cropped up, including a sticking vent valve in the S-IC stage for the liquid oxygen tank.

With the closeout crew retreating across the gantry, the swing arm (No 9) holding it firmly against the spacecraft began to retract to the 12° position, some 10ft (3m) away from the Command Module, at T-43 minutes. From that point on the LES was armed. Just prior to that event taking place, Ken Mattingly entered Mission Control in Houston's Manned Spacecraft Center and joined astronauts John Young and Joe Kerwin at the Capcom console. It was from here that the sole voice to link Earth with Apollo 13 would maintain voice communications. Milt Windler glanced across as Mattingly passed by. 'Sorry to see you here, Ken,' he said with some degree of sincerity.

By T-14 minutes Apollo spacecraft AS-508 (Apollo Saturn 508, the eighth Saturn V to fly) was on full internal power, no longer sharing electrical energy from the fuel cells with a ground supply. With numerous checks and pre-launch sequences cascading down through electrical circuits in the giant rocket, around the world the global tracking network gave sequential 'goes' for launch. Only a non-essential radar tracking ship, the *Vanguard*, was inoperable. From Houston, mission director Chet Lee gave his authority to launch, as did Walt Kapryan, the launch operations director at the Kennedy Space Center.

Scores of engineers in Firing Room 1 watched redline values, ready to intervene and stop the count at any point. Referring to their destination in the undulating features of the Fra Mauro region on the Moon, launch operations manager Paul Donnelly gave his final word to the crew: 'Good luck; head for the hills!' At T-45 seconds Jim Lovell performed the last crew function before lift-off when he set the final guidance alignment in the Command Module Computer.

Leaving Earth

To the spectator watching from a safe distance, the sequence of events associated with the launch of a Saturn V are no different to that of any other rocket: light; smoke, vibration through the ground; noise. In that order. But in the case of a Saturn V it is unimaginably more dramatic – bigger in every way; something totally different to anything experienced with any other rocket that has ever flown.

The power and the noise are indescribable and the all-pervading sense of being overwhelmed is profound, as though the brightness will outdo the Sun. The thunder

coming up through the ground is as encroaching as an earthquake and evokes a sense that the noise will get louder in some violently uncontrolled crescendo threatening to burst ear drums, shock waves pounding on the human chest like thunder from some giant drum.

At lift-off the giant rocket rises gently at first, leaning over away from the launch tower so as not to collide with the swing-arms. That can be disconcerting. Then, 12 seconds after lift-off, the whole stack rolls round to put itself in the proper heading; although a cylinder without wings, the rocket has an up, down, left and right aligned with the orientation of the crew, facing forward in their couches. AS-508 steered to a heading of 72.043° east of north and climbed towards orbit. The effect of this combined pitch and roll, which lasted 20 seconds, was that of a corkscrew manoeuvre until the vehicle was correctly aligned.

As the Saturn climbs ever higher the ride smoothes out for a while, but not for long as vibrations build up again, the giant 36-storey stack continuing to pitch over in a gentle arc. Not always smooth, sometimes pitching and twitching as the giant F-1 engines gimbal on electronic commands; sometimes imperceptibly bouncing up and down, the entire vehicle as though on a trampoline. The crew were alert to this, indications of the dreaded 'pogo' effect.

The vehicle passed through the speed of sound (Mach 1) just 68 seconds off the pad at an altitude of five miles (8km), maximum aerodynamic pressure coming 13 seconds later at a height of 7.8 miles (12.5km). Accelerations were building, but only slowly – 2g at 90 seconds, and 3g at 2min 15sec when the centre F-1 engine shut down on command from the Instrument Unit as planned. The crew sensed a noticeable drop-off in acceleration as the Saturn's thrust was chopped by 20%, rapidly building again to a maximum 3.8g within a mere 28 seconds. At that point the four outer engines shut down on receipt of a liquid oxygen low-level sensor reading, 2min 44sec after lift-off.

The vehicle was already at an altitude of 27 miles (43.5km), 28 miles (45km) from KSC and moving at a speed of 4,314mph (6,943kph). When the S-IC stage separated and fell away just one second after shutdown, eight solid propellant retro-rockets – located in pairs at the

BELOW Thunder across the Kennedy Space Center as the eighth Saturn V climbs slowly into a clear early afternoon sky on Saturday, 11 April 1970. *(NASA-KASC)*

four fin housings down at the base of the stage – fired for a brief 0.6sec. They delivered a thrust of 703,200lb (3,128kN), producing a flash that shot all the way up the side of the Saturn and was visible to the crew.

On the crew display panels, a yellow 'S-II sep' (second stage separation) light came on. In just 164 seconds, the Saturn V and its Apollo payload had shed 77% of their weight at launch. The spent S-IC stage continued on a ballistic trajectory as it arched over, falling into the Atlantic Ocean 409 miles (658km) from KSC, north-east of the Bahamas, 9min 7sec after lift-off. Just under four seconds after stage separation, the five liquid hydrogen/liquid oxygen J-2 engines in the S-II stage lit up, 2min 48sec after lift-off, this action confirmed by five solid lights on the Apollo display console.

To prevent weightless propellants floating in the delivery lines, four 'ullage' motors on the bottom of the S-II stage, attached to an inter-stage structure that would soon be jettisoned, fired for four seconds to provide positive acceleration. These Flexadyne solid-propellant motors each delivered a thrust of 22,500lb (100kN) in a compound designed for high performance and great mechanical strength under the conditions in which they operated.

At separation from the S-IC the astronauts felt a few seconds of weightlessness, difficult to sense other than through loose items floating free. But when the J-2s lit up the acceleration started building once again, in a ride that this time was initially smooth as silk. Cryogenic rocket motors impose a more benign flight without the rumbling vibrations of the S-IC. Just 26 seconds later the circular 11,465lb (5,200kg) inter-stage adapter, 33ft (10m) in diameter and just over 18ft (5.5m) tall, was jettisoned, being a structure that separates the two stages when joined together but was now unnecessary and so much dead weight. And at that moment, aboard Apollo 13, the yellow 'S-II sep' light went out – as it should.

Just under seven seconds after inter-stage separation, the no longer needed Launch Escape Tower fired its jettison motor. Producing a thrust of 31,500lb (140.1kN) for one second, the solid-propellant rocket pulled the tower and the Boost Protective Cover away from the top of the Apollo spacecraft, exposing all

its windows for the first time. The BPC was the 700lb (317kg) cone-shaped covering that protected the exterior of the Command Module from the rocket motors of the Launch Escape System, should they be needed for lifting Apollo away from the Saturn rocket. Combined, the LES and the BPC weighed approximately 8,700lb (3,946kg).

But then things began to go badly wrong. Oscillations began building up and the spacecraft started to vibrate at a frequency of 16Hz and 0.1g. There was nothing really unusual in this. But down below at the S-II crossbeam supporting the centre (No 5) engine, oscillations were at 34g and rising, causing the centre J-2 engine to shut down at 7min 43sec, some 2min 12sec earlier than planned and 3,876mph (6,236kph) short of the planned velocity at CECO (centre engine cut-off).

High amplitude pressure oscillations of about 236lb/in² (1,627kPa) in the liquid oxygen feed lines had activated a switch cutting engine No 5. Had it not shut down, these vibrations could have built up to a destructive level, triggering an in-flight abort and the end of Apollo 13's mission right there. With thrust reduced by 20% it was left to the remaining four J-2 engines to push on, continuing to burn 34.5 seconds longer than would have been necessary with five good engines.

At S-II shutdown, 9min 53sec after launch, the vehicle was 117 miles (189km) above the Earth, 1,110 miles (1,786km) down range and moving at 15,416mph (24,810kph) almost parallel to the surface far below. It had made up all but 149mph (240kph) of the velocity deficit. Apart from the vibrations, the long ride on the S-II had gradually increased the rate of acceleration, peaking at 1.7g just before OECO (outer engine cut-off).

One second after OECO the S-II separated, fired up its four retro-rockets to prevent it bumping into the third stage, and fell away. The S-II broke up as it fell back through the atmosphere, falling into the Atlantic 2,822 miles (4,542km) from the Cape, south-west of the Azores, 20min 58sec after launch. A very different fate lay in store for the S-IVB (the Saturn rocket's third stage), which ignited its single J-2 engine exactly one second after separation. Two solid-propellant Thiokol TX-280 rocket motors provided a thrust of 3,390lb

RIGHT Events planned for the mission were laid out in a Flight Plan comprising several hundred pages, of which this page from the author's personal copy covers the 60 minutes from 03:00 GET (ground elapsed time) to 04:00 GET. The standard pro forma layout carried a column on the extreme left for Mission Control-Houston (MCC-H), with the elapsed time, next right, running from top to bottom. The dotted line next right shows when the TV transmission was scheduled, followed by the time the spacecraft was in communication with the Manned Space Flight Network (MSFN). The main bulk of activity occupies the centre block of information, with separation of *Aquarius* from *Odyssey* and docking marked.
(David Baker)

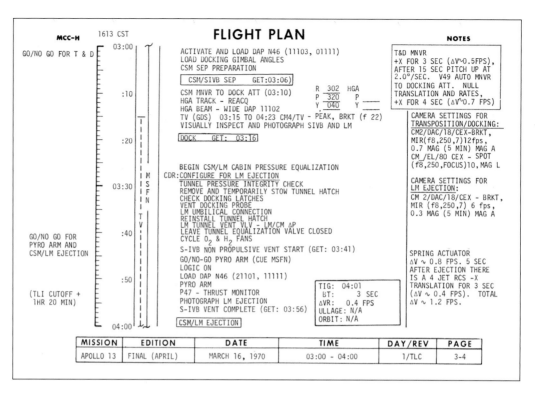

(15.08kN) for four seconds. These 'ullage' motors fired just before ignition of the J-2 after S-II stage separation to provide positive acceleration and settle the propellants in the bottom of the main tanks for a clean engine start.

The S-IVB gave the crew a gentle 0.7g at maximum acceleration as the single J-2 shut down, nine seconds later than planned to compensate for the under performance of the S-II. When the S-IVB's J-2 stopped firing 12min 30sec after launch the stack was 117 miles (192km) above the Earth, 1,765 miles (2,840km) down range and travelling at a speed of 17,429mph (28,049kph). Remarkably, for all the irregularities encountered during ascent, the Saturn V had delivered the S-IVB and Apollo complex to orbit within 1.3mph (2.1kph) of the pre-planned velocity, but not in quite the right orbit.

The resulting orbit was 112.8 miles (181.5km) by 115 miles (185.5km) with a 1.9ft/sec (0.6m/sec), or 1.3mph (2.1kph), under-speed and a heading angle of 1.2° greater than the target value. Also, the early CECO on the S-II had a compounding effect in reducing the margin of error in the S-IVB stage for TLI (trans-lunar injection) when the S-IVB would fire up a second time to send Apollo out of Earth orbit and on its way to the Moon.

There had to be a certain quantity of propellant

in the S-IVB to achieve that. In burning longer to reach Earth orbit, the S-IVB had used up propellant that would normally have served as a margin between the quantity required to leave Earth orbit and the amount held in the tanks. Only 295ft/sec (90m/sec), or 216mph (348kph), of propulsive capability would remain in the S-IVB tanks after a nominal TLI burn, half the normal quantity. TLI was scheduled to take place toward the end of the second orbit, after comprehensive checks had determined the precise time at which to fire the S-IVB.

TLI

For the crew, the period in Earth orbit was important for conducting initial checks of the Inertial Measurement Unit (IMU) using the stars Spica and Antares, and for aligning the guidance platform. It was a busy time, and as the spacecraft drifted round the Earth, in and out of the ground-based tracking stations, Capcom Joe Kerwin kept up an intermittent dialogue. In an otherwise flawless sequence, the only problem was that the dust covers failed to jettison when the shaft was driven, but did so when the optics were being driven to the first star.

The Crew Optical Alignment Sight (COAS) was unstowed and set up in the left forward-facing window and the horizon check was successfully completed. The COAS is a small telescope device, 8in (20.3cm) long, attached to a mount on a window frame, which presents an image of an alignment target on *Aquarius* when *Odyssey* approaches it for docking after TLI. Next, the probe on the front of *Odyssey* was extended, ready to snag on to the docking cone of *Aquarius* shortly after leaving Earth orbit.

Then it was time for the first TV show, which had aimed to show the coastline of the Gulf of Mexico, impossible to see due to cloud cover; the six-minute show began at 1hr 38min into the mission (or GET, ground elapsed time). In Mission Control, Windler and his team monitored systems and activity aboard *Odyssey*. Polling the Maroon Team flight controllers he came up 'go' for Trans-Lunar Injection, voiced up to the crew by Joe Kerwin a couple of minutes after the TV was switched off. The spacecraft was crossing the Atlantic Ocean at the end of its first orbit.

Shortly thereafter, losing contact through the Canary Island station, it began a 30-minute period out of contact before appearing over the horizon at the Carnarvon station in Australia. And then, disappearing again over the horizon, the S-IVB began computerised preparation for the TLI burn. Three ARIA (Apollo Range Instrumentation Aircraft) were positioned along the ground track between the stations in Carnarvon and Hawaii. They would seek to maintain communication with the crew and capture telemetry from the S-IVB as it performed the burn.

The ARIA were converted Boeing EC-135 aircraft adapted from the Boeing 367-80. Equipped with comprehensive radar and radio equipment, the eight ARIA, distinctive with their exaggerated bulbous nose chin, were based at Patrick Air Force Base south of Cape Canaveral. Communication with Apollo 13 was scratchy at best, but there was intermittent voice exchange with Joe Kerwin in Mission Control.

Ignition for the TLI burn came at an elapsed time of 2hr 35min 46sec and lasted 5min 51sec. This time, with reduced weight due to diminishing propellant, the ride on the S-IVB was faster than on the first burn climbing into Earth orbit, the acceleration greater this time, building to a peak 1.4g at shutdown. At that

moment, the stack was moving at 24,230mph (38,986kph) at a flight path angle of 7.6° to the surface of the Earth. The crew reported a vibration during the burn, noted on previous flights also, but were generally impressed with the performance of the engine and the rapid build-up of speed.

This was the fifth time astronauts had departed on missions to the Moon, but only the second time Apollo would be placed on a trajectory from which it could not return to Earth by simply looping around the Moon and coming home. Apollo 12 had pioneered this non-free-return trajectory, as it was known.

On a non-free-return trajectory, a propulsive burn would have to occur to put the spacecraft back on a free-return path. Mission planners were not prepared to fly straight to a non-free-return trajectory before a back-up engine – the LM Descent Engine (LMDE) – had been made available, so that if the Apollo SPS engine became inoperable the LMDE could do the job. The solution was first to use the S-IVB's TLI burn to put Apollo on a free-return path, and only then conduct a midcourse correction burn with the SPS to place the vehicles on a non-free-return route after the SPS had been shown to work.

The hybrid trajectory began when the S-IVB shut down, aiming for a pericynthion (low pass around the Moon) of 478.5 miles (770km). The mission plan included a possible midcourse correction (MCC-1) at about nine hours after TLI. This was intended to correct any deviations in the actual trajectory imparted by the S-IVB, but these were so negligible as to be unnecessary. MCC-2, however, at around 30hr 40min GET, would alter the trajectory to a non-free-return path with pericynthion at around 69 miles (111km).

Within three hours of TLI, Apollo 13 would be only 25,000 miles (40,225km) from Earth but the planet's gravity, diminishing all the while on the inverse square law, would slow the spacecraft to 8,200mph (13,194kph) – a third of the velocity imposed by the S-IVB stage. Three hours after that, it would have slowed to 6,750mph (10,860kph) and the spacecraft would have travelled 49,000 miles (78,840km) from Earth. It would continue to slow as it drifted towards the Moon and eventually come within the Moon's gravitational grasp, when it would start to build up speed once more.

RIGHT The docking system for Apollo consisted of a probe on the Command Module side, a drogue in the Lunar Module, and a latching system on the docking ring to hold the two spacecraft rigidly together. *(North American Rockwell)*

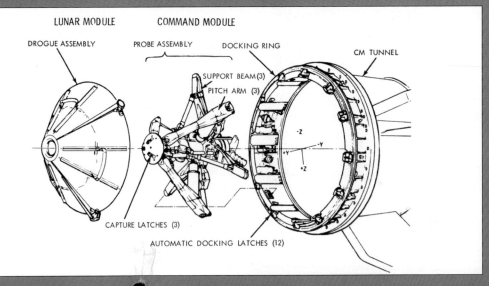

LUNAR MODULE   COMMAND MODULE

DROGUE ASSEMBLY

PROBE ASSEMBLY

DOCKING RING

CM TUNNEL

SUPPORT BEAM (3)

PITCH ARM (3)

-Z

-Y

+Y

+Z

CAPTURE LATCHES (3)

AUTOMATIC DOCKING LATCHES (12)

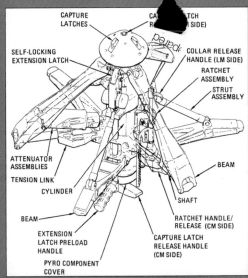

CAPTURE LATCHES

CAPTURE LATCH RELEASE (LM SIDE)

SELF-LOCKING EXTENSION LATCH

COLLAR RELEASE HANDLE (LM SIDE)

RATCHET ASSEMBLY

STRUT ASSEMBLY

ATTENUATOR ASSEMBLIES

TENSION LINK

CYLINDER

BEAM

SHAFT

BEAM

RATCHET HANDLE/ RELEASE (CM SIDE)

EXTENSION LATCH PRELOAD HANDLE

CAPTURE LATCH RELEASE HANDLE (CM SIDE)

PYRO COMPONENT COVER

BELOW The spring-mounted attenuator assemblies (covered with aluminium sleeves for flight) provide shock absorption by acting against the conical wall of the drogue. *(NASA)*

LEFT With aluminium inner and outer cylinders, the docking probe has an extension capability of 10in (25.4cm) and a self-centering, gimbal-mounted probe head with three capture latches. The retraction mechanism consists of a cold-gas system pressurised by four nitrogen bottles inside the body of the probe. *(North American Rockwell)*

BELOW The probe head contains the three capture latches which engage automatically when they snag the central receptacle on the conical drogue. This is known as 'soft dock', a 'hard dock' being achieved when the automatic retraction mechanism is triggered by microdots in the capture latches firing the nitrogen cylinders. *(North American Rockwell)*

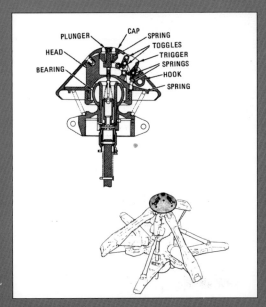

PLUNGER

CAP

SPRING

TOGGLES

HEAD

TRIGGER

BEARING

SPRINGS

HOOK

SPRING

DOCKING

The docking device consisted of two halves, the retractable probe installed in the forward section of the Command Module and the 31.5in (80cm) conical drogue fixed to the top of the Ascent Stage of the Lunar Module. Bolted to the forward section of the tunnel on the Command Module above the circular forward hatch, the aluminium ring incorporated 12 latches spaced at equal intervals around the circumference. When sprung on contact with the docking flange on the LM, it ensured a rigid structural and pressure-tight connection. When docked the two vehicles could be manoeuvred as a single structure.

The docking probe comprised an inner and outer cylinder capable of extension or retraction to a maximum of 10in (25.4cm), supported by three arms fixed to the docking ring at 120° intervals. The forward end of the inner cylinder supported a cone with three small capture latches spaced at 120° intervals. These were designed to engage with a circular opening in the drogue and to maintain a grip on the cone as the probe cylinder was retracted, pulling the two vehicles together for 'hard dock', when the docking latches engaged with the rim of the docking flange on the LM. Three impact attenuators absorbed the energy on contact with the drogue.

ABOVE The extended probe set within the docking ring. *(NASA)*

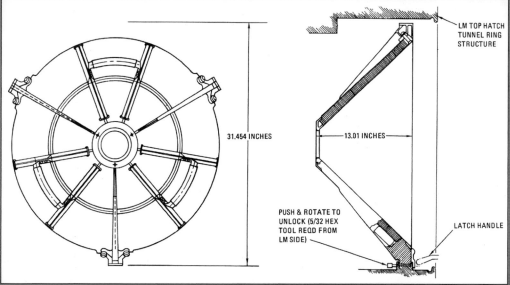

LEFT Fabricated from aluminium honeycomb with aluminium support beams, the drogue assembly has a locking system to prevent it rotating during docking. *(North American Rockwell)*

RIGHT The aluminium docking ring is bolted to the top of the Command Module just forward of the top hatch, and is the mounting point for the probe and the 12 docking latches, any three of which will secure an air-tight seal. The latches spring on contact and grip the rear surface of the LM docking flange. The ring has a pyrotechnic charge for physical ejection with the LM at the end of a lunar mission. *(North American Rockwell)*

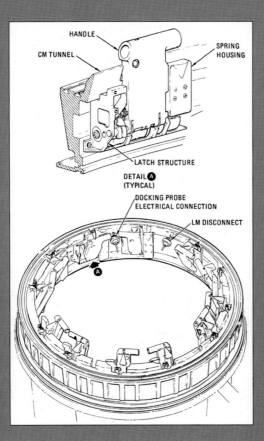

The removable probe assembly was designed to collapse so that it could be taken out of the docking tunnel and stowed, allowing passage for crew members from the Command Module through the short tunnel into the Lunar Module. The docking tunnel in which the probe and drogue assembly was located was sealed on the Command Module side by a circular hatch. This could be completely removed and stowed when the tunnel itself had been pressurised, allowing access to the docking equipment.

After checking that all the docking latches had sprung, and manually locking any that had not engaged, the probe assembly was collapsed and removed from the tunnel. When the Ascent Stage was pressurised the LM hatch could be opened from the CM side, which hinged back down into the Ascent Stage to an angle of 70°, not quite at right angles to the closed configuration.

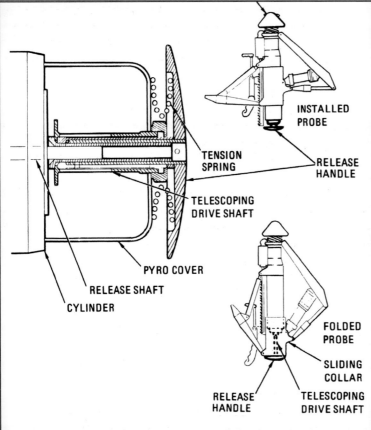

LEFT After docking with *Aquarius*, pressure equalisation and opening of the top hatch in *Odyssey*, the probe is folded and the three capture latches released using a plunger on the base. *(North American Rockwell)*

BELOW For reinstalling the probe in the docking ring, a ratchet tool is carried aboard *Odyssey* for manually installing the probe prior to sealing up the tunnel. *(North American Rockwell)*

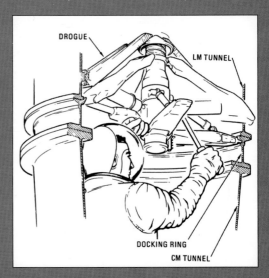

TD&E

Aboard Apollo 13 Swigert took up position on the left couch, with Lovell in the centre and Haise in his usual position on the right couch. The spacecraft had to separate from the SLA (Spacecraft Lunar Module Adapter) to a distance of about 150ft (45.7m), pitch 180° so that it was facing the S-IVB/SLA, roll 60° and translate forward to the Lunar Module, nestled inside the bottom section of the SLA. This operation, known as Transposition, Docking and Extraction (TD&E) began with separation of *Odyssey* from the SLA at 3hr 6min 39sec GET, just 25 minutes after TLI.

Apollo spacecraft were attached to the top of the four detachable SLA panels by a flange around the entire circumference of the two structures, with an explosive charge to cut through the connection. Two strands of redundant detonating cord with 28 redundant charge holders ensured clean separation when fired, operating like a high-speed fuse to sever the two structures.

A combination of pyrotechnic thrusters, springs and pistons rotated the four detachable panels past an angle of 45°, at which point they separated from the fixed lower section and were ejected, separating at an angle of 110° and a speed of almost 4mph (6.4kph), never to be seen

again. When *Odyssey* separated from the SLA the loud bang as the pyrotechnic fuse severed the two structures was a shock if unexpected.

Once again, aboard *Odyssey* it was time to get out the TV camera and give Earthbound viewers an astronaut's eye view of the S-IVB and of the docking with *Aquarius*. With Jack Swigert in the left couch manoeuvring the spacecraft,

SUPPORT

PANEL

UPPER HINGE

UPPER HINGE

LOWER HINGE

LOWER HINGE

SPRING THRUSTER AFTER PANEL DEPLOYMENT, AT START OF JETTISON

SPRING THRUSTER BEFORE PANEL DEPLOYMENT

LEFT The separation mechanism for ejecting the panels of the Spacecraft Lunar Module Adapter (SLA) was added when on Apollo 7 the crew advised that retaining the hinged panels could pose a collision hazard to later missions when LM extraction was needed. *(North American Rockwell)*

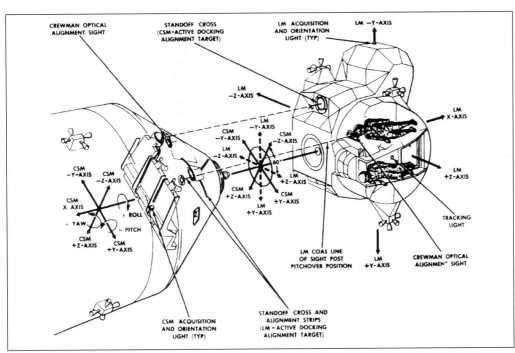

LEFT Although *Aquarius* was unmanned when extracted from the top of the S-IVB stage, this crew-training diagram shows the relative axial orientations of *Odyssey* and the Lunar Module for docking purposes. *(North American Rockwell)*

RIGHT The Crewman Optical Alignment Sight (COAS) consisted of a collimator-type device providing the user with a fixed line-of-sight reference image. For rendezvous with *Aquarius*, Jack Swigert installed it in a mount attached to the forward No 2 window in the Command Module as he occupied the left couch. When viewed through the COAS, the image appeared to be at the same range as *Aquarius*. A light adjustment allowed for brightness against the background. Range and rate of closure were obtained by the relative size of the image and the target. *(North American Rockwell)*

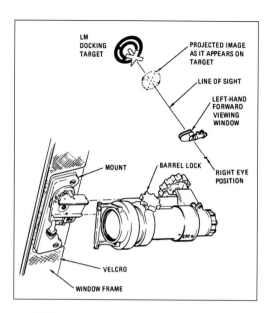

Fred Haise operated the camera from his right seat. Swigert would 'pilot' *Odyssey* to a docking using one of the two forward-facing rendezvous windows. Using the COAS to get a bead on the docking target atop *Aquarius*, Swigert squeezed his hand controller and edged *Odyssey* toward the exposed Lunar Module.

There were two rotation control sticks in the Command Module, giving the CMP the closest experience of 'flying' a spacecraft it was possible to get. Each comprised a square box with a three-axis attitude control stick, 5in (12.7cm) long. For manoeuvring *Odyssey* in pitch, roll or yaw the hand controller was used with the box locked on a bracket at the end of the armrest. For convenience there was a push-to-talk button on the handle.

For TD&E, Swigert had the translation controller on his left armrest and the rotation controller on his right rest. Movement forward, backward, up or down would be achieved through the T-handle on his left armrest. Spinning the spacecraft around would be achieved with the joystick at his right hand.

To move the spacecraft forward away from the SLA, Swigert pushed in the T-handle on the translation controller to operate the thrusters that pushed the spacecraft forward, pulling it back out to use braking thrusters and stop the relative motion when he was sufficiently far away. Because he could not see how far he was moving ahead of the S-IVB and *Aquarius*, he used the Entry Monitor System (EMS) to watch the rate of separation.

The EMS (which, for all the different jobs

BELOW The COAS could be used from either *Odyssey* or *Aquarius,* and this diagram shows how that would facilitate alignment and docking operations using a tee-cross docking target on each spacecraft. In the case of Apollo 13, only the one docking was conducted, when *Aquarius* was extracted from the SLA mounting. *(North American Rockwell)*

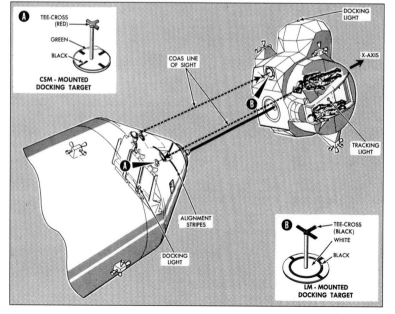

BELOW A ground-based alignment check of docking probe and drogue between a Command Module and a Lunar Module Ascent Stage depicts the way the two spacecraft are connected for the trans-lunar journey. *(NASA)*

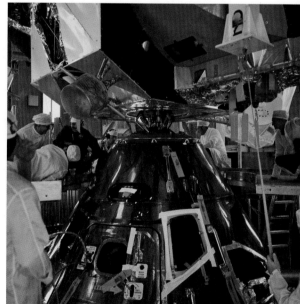

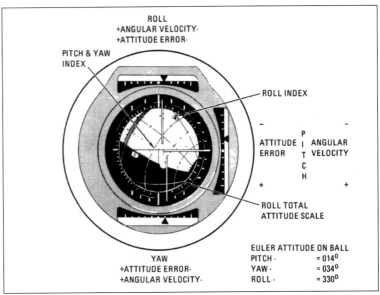

ABOVE **Key to maintaining a known orientation against a fixed frame of reference was the Flight Director Attitude Indicator (FDAI), also known colloquially as the 'eight-ball'. There were two in the Command Module, each 6.9in (17.5cm) in diameter and 9.3in (23.6cm) long, weighing 9lb (4.1kg) with glass cover plate.** *(MIT)*

it performs, could just as easily be called the 'energy management system') is a screen located on the main display panel and is supposed to give velocity data, but it can be erratic when a lot of other vehicle motions are in play. Nevertheless, the ideal is to see a forward motion of 0.5–0.8ft/sec (0.15–0.24m/sec) and halt the drift at that point.

Crews varied in their enthusiasm for relying on the EMS and some were openly hostile to its use, preferring to blip the T-handle translation controller and wait the appropriate number of seconds to achieve the required separation, 'in the blind'. Swigert held the thrusters open for 4.3 seconds, imparting a change in velocity (ΔV, or delta-v) of 0.9ft/sec (0.27m/sec), before halting the forward motion.

To turn around and face *Aquarius*, Swigert used the rotation controller with his right hand to command pitch-up by pushing the stick back, reversing the action to bring *Odyssey* to a halt when it had turned through 180°. Greeted with a snowstorm of debris cast off when the CSM separated from the SLA, the crew

could see the jettisoned panels floating away, destined to orbit the Sun forever.

With attitude now stable, Swigert pushed forward the T-handle translation controller and propelled *Odyssey* back toward *Aquarius*, using the COAS to get the best visual focus on his target, reversing the T-handle action to slow the rate of closure. With practice it is possible to get the contact speed down to a rate of 1in/sec (2.54cm/sec), but that takes a lot of work, and a very slow rate of closure is not really desirable in docking: the docking system needs firm recognition of a positive contact to spring the latches that will hold the two spacecraft together, and 'kissing' the collar is not guaranteed to get a reaction.

Swigert was proving to be a competent pilot and performed a flawless docking, making contact with *Aquarius* at 3hr 19min 9sec Ground Elapsed Time (3:19:09 GET). First with the three small capture-latches in the head of the extended probe, the two vehicles – just a few inches apart, connected by only the three tiny latches – had a 'soft dock'. As the probe seeks the receptacle in the drogue some swaying motion can occasionally be sensed, but there was none in *Odyssey*, and Swigert drove the CSM forward, making contact at 0.2ft/sec (0.12mph or 0.2kph). Then it was time to retract the probe's inner cylinder and draw the two vehicles into 'hard dock' position, announced by a very audible 'clunk' indicating the docking latches had fired to the closed position.

ABOVE **The FDAI is a ball with circles showing spacecraft attitude in pitch, roll and yaw. Three scales, above, to the right and below the ball, show pitch, roll and yaw rates and a needle shows error in each axis. The FDAI receives signals from the stabilisation and control system or the guidance and navigation system, while rate displays originate in the stabilisation and control system only.** *(MIT)*

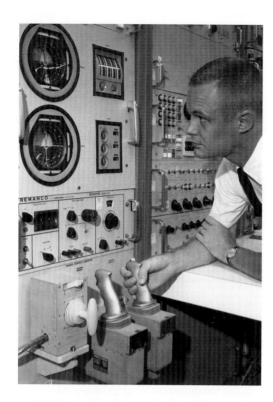

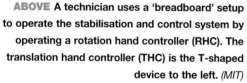

ABOVE A technician uses a 'breadboard' setup to operate the stabilisation and control system by operating a rotation hand controller (RHC). The translation hand controller (THC) is the T-shaped device to the left. *(MIT)*

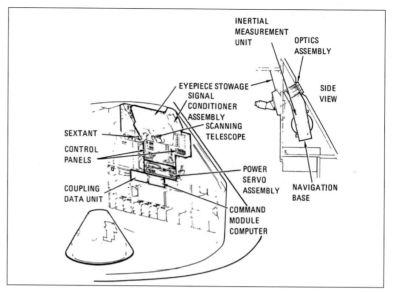

ABOVE The optical subsystems at the guidance and navigation station in the lower equipment bay are integral with the fixed navigation base and its inertial measurement unit. *(MIT)*

As Jim and Jack went through the post-docking checklist, Fred Haise panned the TV camera to show the scenes inside *Odyssey*, working his way up toward the tunnel connecting the Command Module to the LM. Several procedures were now essential to confirming the validity of the docked configuration before separating from the top of the S-IVB. Lovell scurried up into the short tunnel between the two hatches to check pressure equalisation between

RIGHT The main components of the guidance and navigation equipment are situated in the lower equipment bay, where the optical devices, sextant and telescope, are mounted for use by the crew in taking navigational sightings. Note the optional mounting for one of two rotation hand controllers carried in *Odyssey* so that an astronaut can 'turn' the spacecraft around for sighting the optics on star or landmark sightings. *(North American Rockwell)*

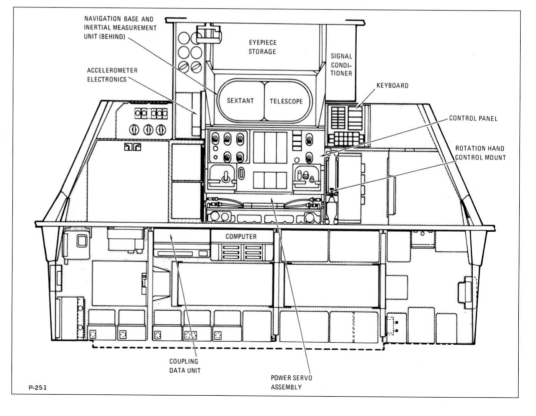

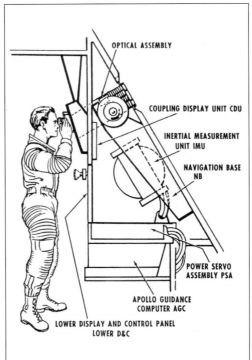

them. Then he removed the Command Module forward hatch and checked the docking latches to discover that two of the 12 (Nos 1 and 4) had not engaged.

Lovell connected umbilicals between *Odyssey* and the LM and electrical power was switched on in *Aquarius* at 3hr 46min GET. This allowed Mission Control to see current levels – an indication of status – as Swigert got ready to extract the LM and pull it free from the lower section of the SLA.

While all this was going on, Apollo 13 began to slip through the Van Allen radiation belts, the charged fields of radiation trapped in Earth's magnetosphere and the most intensive radiation environment of the mission. Much mischief has been made about the dangers of exposure to this radiation, which was recorded at a maximum skin dose rate of 2.27rad/hr and a maximum depth dose rate of 1.35rad/hr. Conspiracy theorists claiming humans could not survive the belts fail to acknowledge that Apollo flew up and over the tear-drop shaped belts and not directly out through the zones of maximum depth or intensity.

Docked together, the CSM/LM combination separated at precisely 04:01 GET. This was achieved when Swigert lifted up the guard over the 'S-IVB/LM Sep Switch' mounted just below

the DSKY in front of his couch. Finding it stiffer to operate than in the simulator back on Earth, he threw the switch that fired pyrotechnic squibs severing the tie-down links at each of the four landing gear outriggers on the Descent Stage. Spring actuators pushed the docked spacecraft away at a speed of 0.8ft/sec (0.24m/sec).

From the left seat Swigert blipped the T-handle translation controller and pushed that up to 1.2ft/sec (0.37m/sec) as the docked combination of *Odyssey* and *Aquarius* backed away. In all, for TD&E, Jack had used a relatively frugal 55lb (25kg) of thruster propellant.

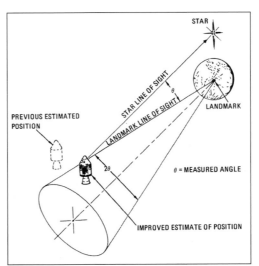

RIGHT Several alignments are made throughout the mission, ideally at times prior to major manoeuvres involving a course correction or a change in the flight path. At these times either Earth or the Moon are used for alignments. *(North American Rockwell)*

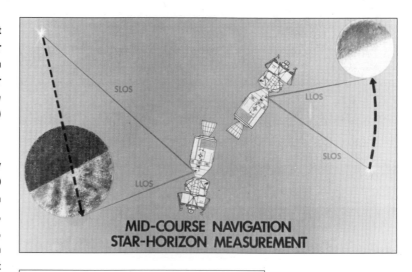

CENTRE The display and keyboard assembly (DSKY) consisted of ten numerical keys labelled 0 to 9, two sign keys (+ or –) and seven instruction keys (verb, noun, clear, proceed, key release, enter and reset). When a key is depressed, 14 volts are applied to a diode encoder which generates a five-bit code associated with that key. A noun code refers to a device, or a group of computer registers (the operand), while a verb code indicates what action is to be taken (the operator). *(North American Rockwell)*

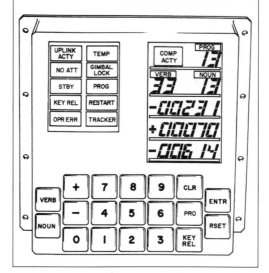

BELOW Produced by the AC Electronics division of General Motors, the Inertial Measurement Unit (IMU) was a ball-shaped unit with a diameter of 12.6in (32cm) weighing 42.5lb (19.3kg). The inner gimbal was the stable member to which was mounted three gyroscopes and three accelerometers. The gimbals are connected to each other by drive motors and angle resolvers, and the unit usually requires 217 watts of power at 28 volts DC. *(MIT)*

BELOW World famous for its inertial platforms dating back to the Second World War, Sperry Gyroscope made the matched pendulum and module assembly for the IMU. *(Sperry)*

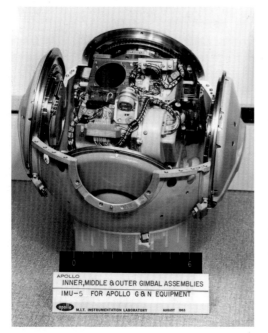

HAND CONTROLLERS

Each Lunar Module crew member has an Attitude Controller Assembly (ACA) operated by his right hand and a Thrust/Translation Controller Assembly (TTCA) for operation with his left hand.

The ACA controlled roll, pitch and yaw via a pistol-grip, with hand movements analogous to vehicle rotations around these axes. Pitch was controlled by forward or aft tilt of the hand grip, with roll right or left controlled by tilting the ACA to left or right, yaw induced by rotation around the vertical. Transducers in the unit produced attitude rate commands proportional to the degree of displacement from the neutral position, with limit switches wired to the RCS thrusters.

To physically move the LM about and to change its orbit, the TTCA was used to fire the RCS thrusters and to send throttle commands to the Lunar Module Descent Engine via a control unit.

A lever on the right side allowed the crew member to select one of two control options: translation in the Y and Z axes using the RCS thrusters and LMDE throttle commands for X-axis translation combined, or translation in all three axes using the thrusters alone. As with the ACA, hand movements conformed to axial movement of the LM but in physical translation rather than attitude control, and the mode selected was controlled by a level set to either the up or the down position.

For translation using thrusters for Y and X axes and thrust level control of the main engine for Z-axis motion, the lever was set to the up position and the T-handle tilted up or down for increases or decreases in throttle settings. With the lever in the down position the RCS thrusters controlled translation in all three axes, but in both cases Z-axis translation was controlled by inward or outward motion of the handle.

With the lever in the up position the Y and X axes command movements were spring-loaded to the middle detent position. In this case the main engine throttle setting remained as positioned by the astronaut. When the lever was in the down position, using only thrusters for translation, commands to all three axes were spring-loaded to the neutral position. Under PGNCS control the LGC commanded the engine to fire and to stop at the time entered by the astronaut prior to the burn.

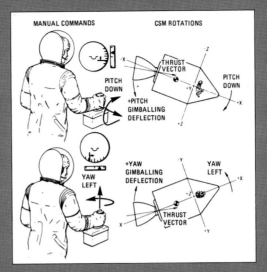

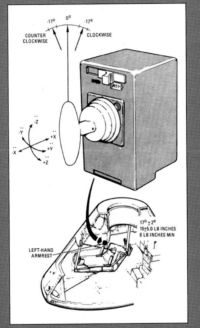

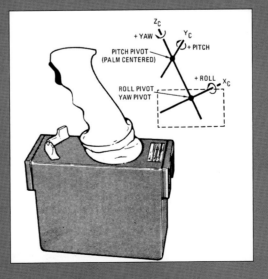

LEFT Thrust-vector control of the Service Propulsion System (SPS) is achieved manually using the Translation Hand Control (THC) incorporating a T-stick. The control is mounted with its axes approximately parallel to those of the spacecraft. *(North American Rockwell)*

LEFT The THC can also be used for manual control of velocity changes with the RCS thrusters by operating it in the manner depicted on this training diagram. The THC can be rotated in either direction about the centreline of the shaft on which it is mounted, with hard stops for each rotation at +/–17° from null, with detent positions at a nominal +/–12°. *(North American Rockwell)*

BELOW The Rotation Hand Controller (RHC) is used for changing the attitude of the spacecraft, or of the docked configuration, by sending commands through the stabilisation and control system to the RCS thrusters. Two are carried in *Odyssey* and each has a tapered female dovetail at each end of the housing which mates with mounting brackets on the couch armrests and at the navigation station in the lower equipment bay. Transducers produce alternating current proportional to the displacement of the RHC. *(North American Rockwell)*

Cruising to the Moon

Its primary role completed, the S-IVB had maintained a stable attitude for TD&E. Eight minutes after the LM was extracted the S-IVB began to manoeuvre for an 80-second burn with its two auxiliary propulsion system (APS) thruster modules to separate from the docked CSM/LM, that event taking place nine minutes later. This was followed at 04:39:19 GET with a planned liquid oxygen dump. The initial free-return path had set the stack on course for a fly-by around the far side of the Moon. But now that *Odyssey* and *Aquarius* had docked and pulled away from the third stage, the S-IVB would be redirected for an impact with the Moon.

Now that a seismometer powered by a nuclear power source had been installed on the surface by the Apollo 12 astronauts, the spent stage could be used to set up shock waves. Geologists could use those vibrations to obtain information about the internal structure of the Moon. The APS modules fired for 217 seconds six hours into the mission, producing a ΔV of 28ft/sec (8.5m/sec), sufficient to pull it round from a fly-by of the Moon to an impact at the assigned spot on the surface.

Back on Earth, a shift change was under way as Gold Team flight controllers, headed by Gerald Griffin, replaced Milton Windler's Maroon Team, and Capcom Vance Brand replaced Joe Kerwin. In space, an Earth Weather Photography experiment started at precisely 7:17:14 GET. The plan was to take one photograph every 20 minutes for three hours, and Jim Lovell took charge of that task.

It was a test to see if very high altitude photography of the Earth could help meteorologists measure the height of clouds above the surface. The last of 11 such images was taken at 11:37:19 GET at an altitude of 65,283 miles (105,040km) about 4hr 20min after the first. After the photographs were returned to Earth, meteorologists confirmed the theory to be proven. Then it was time for Jack Swigert to set the docked vehicles slowly spinning in a manoeuvre known as PTC (Passive Thermal Control) – vital for controlling the amount of heat flowing to the interior of *Odyssey* and *Aquarius*.

RIGHT The inter-relationship between the communication, environmental, electrical, guidance, stabilisation and propulsion systems was integral to the operating simplicity of the Apollo spacecraft. Highly complex in design and engineering, the integrated systems and subsystems were a hard learn for astronauts, but entirely logical in layout, design and operation. When mastered, the Apollo spacecraft could be made to dance and sing, with infinite adaptability, flexibility and robust response. *(North American Rockwell)*

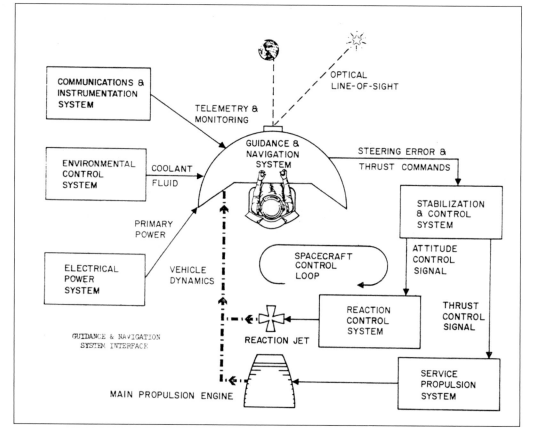

In space, the temperature range can exceed 260°C (500°F). Apollo had a thermal control system using water glycol coolant fed through pipes and rails to maintain an even temperature for electronic systems and propellant tanks. Heat removed from the spacecraft was sent to one of two radiators located on the exterior surface of the Service Module. These comprised aluminium panels, each with a surface area of 4.55m² (49ft²) and a surface arc of 130° around the Service Module. Each panel had five primary coolant tubes carrying the water glycol mixture (62.5% ethylene/37.5% water), and the total system could reject thermal energy at a rate of 4,653kJ/hr (4,415Btu/hr).

Despite this robust capability, to prevent overloading the capabilities of the thermal control system required finding a means of lowering thermal exposure. The solution was to slowly rotate the spacecraft using a passive mode to control the temperature – hence Passive Thermal Control. To set this up, ideally the long axis of the docked vehicles should be aligned perpendicular to the spacecraft–Sun line and made to slowly rotate at a constant rate.

For this reason crew members referred to it as the 'barbecue mode'. Accordingly, Jack Swigert set up the PTC mode to begin at 7hr 30min by configuring the Digital Auto-Pilot (DAP) to control rotation of the spacecraft about its longitudinal axis at 0.3°/sec.

The DAP is not a piece of hardware, rather a systems concept derived from a decision in early 1964 to incorporate in Apollo, for the first time, what would now be termed a digital fly-by-wire system. This concept was a pioneering endeavour and grew out of experience with a similar system used on the Titan missile. It demonstrated the first transition from manual or computer-controlled stick-and-linkage systems to electronic controls governed by a set of computer program algorithms – the first time it had moved from missile to man-carrying spacecraft.

Prior to the decision to adopt the DAP concept, the Apollo spacecraft was being designed with two separate and independent guidance, navigation and control systems. One used a computer built by MIT's Instrumentation Laboratory for guidance and navigation. It

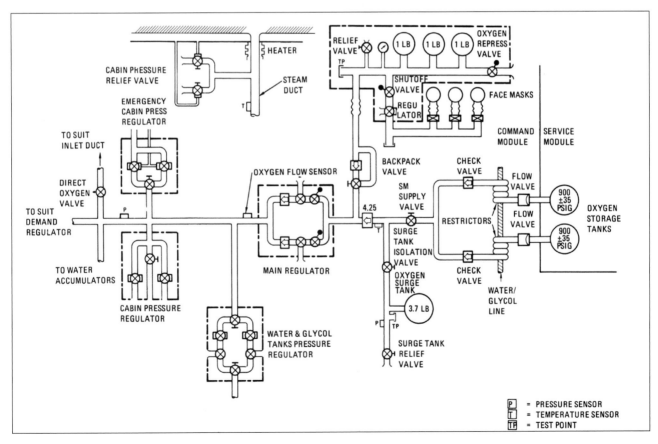

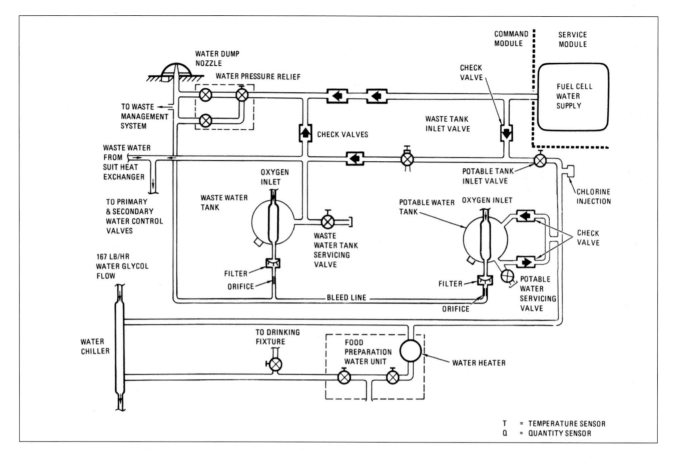

was known as the Apollo Guidance Computer (AGC). Another controlled the attitude and orientation of the spacecraft and was the responsibility of Minneapolis-Honeywell. They were independent systems with their own duplicated sets of hardware and wiring supporting a range of tasks that were only effective when integrated.

When Joseph Shea became head of NASA's Apollo Spacecraft Program Office in October 1963 he disliked this split between two systems using the same hardware. He supported Apollo manager Clifford Duncan in giving MIT the job of doing the work of both computers, integrating functions into a digital autopilot capability through one computer – the AGC. The Honeywell system was split off as an analogue back-up. This functional change was to be the standard for all Block II spacecraft.

The AGC and the DAP function were integral to the operation of the Apollo spacecraft, and so drastic was this compared to earlier manned space vehicles such as Mercury and Gemini that some astronauts threatened to resign. Nobody had built anything operational

before which used digital fly-by-wire, and to relegate that function to a computer was a step too far for some pilots – especially when, only a decade earlier, 'computers' were women working hand-driven machines doing calculations in design offices and test centres! However, the world's first fly-by-wire flying machine, the Lunar Landing Research Vehicle, was taken to the air for its inaugural flight on 30 October 1964 by test pilot Joe Walker.

With this machine, vital for training astronauts how to control vertical thrust for Moon landings, prejudice evaporated. The astronauts found it challenging and that was OK. Once the systems were up and running the astronauts soon found themselves test pilots of a new concept – flying and evaluating electronic control systems rather than new types of hardware with rocket motors.

So it was that when Jack Swigert set up the DAP for PTC, he dialled in a rotation rate of 0.3°/sec and assumed the fly-by-wire system would put the Apollo/LM configuration in a slow barbecue roll evenly distributing solar heat around the exterior. It did not. When the Gold Team GNC (guidance, navigation and control officer)

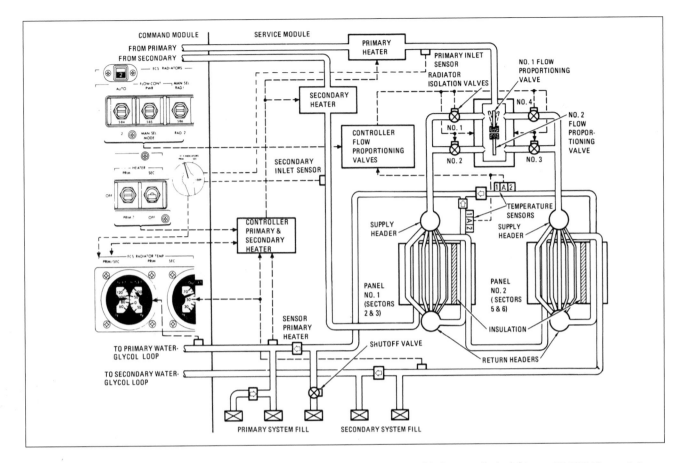

COMMAND MODULE   SERVICE MODULE

FROM PRIMARY
FROM SECONDARY

ECS RADIATORS

FLOW CONT   MAN SEL
PWR   RAD 1
AUTO

MAN SEL MODE   RAD 1

HEATER
PRIM   SEC

PRIM 2   OFF

ECS RADIATOR TEMP
PRIM/SEC   PRIM   SEC

TO PRIMARY WATER-
GLYCOL LOOP

TO SECONDARY WATER-
GLYCOL LOOP

PRIMARY HEATER

PRIMARY INLET SENSOR

RADIATOR ISOLATION VALVES

SECONDARY HEATER

CONTROLLER FLOW PROPORTIONING VALVES

SECONDARY INLET SENSOR

CONTROLLER PRIMARY & SECONDARY HEATER

NO. 1 FLOW PROPORTIONING VALVE

NO. 1
NO. 4

NO. 2
NO. 3

NO. 2 FLOW PROPORTIONING VALVE

TEMPERATURE SENSORS

SUPPLY HEADER
SUPPLY HEADER

PANEL NO. 1 (SECTORS 2 & 3)

PANEL NO. 2 (SECTORS 5 & 6)

INSULATION

SENSOR PRIMARY HEATER

SHUTOFF VALVE

RETURN HEADERS

PRIMARY SYSTEM FILL   SECONDARY SYSTEM FILL

watched the rates they slowly began to wander off at different rates in various axes. And that was no good when the spacecraft were supposed to be slowly rolling at a rate that would see one complete revolution every 20 minutes, a sedate roll rate of three turns per hour.

The problem was run to ground by engineers in the Mission Evaluation Room (the MER, where teams of specialists provided back-up to the flight controllers from Building 45, a good five-minute walk from Mission Control). They found that a crucial setting had been left off the onboard checklist and flight plan notes. To effect a roll rate of 0.3°/sec, the DAP had to be overloaded with a rate command of 0.375°/sec, with all thrusters enabled in roll.

In Mission Control GNC was adamant that this was the correct setting and the MER people confirmed it. It had been used on Apollo 12, but the 'all thrusters enable' requirement had been a late change written by hand into the Mission Control checklists but not those taken aboard *Odyssey*. And Mission Control admitted to an error when Vance Brand acknowledged the mistake: 'We might have sent a confusing

bit of info up to you...we didn't mean that...' At 08:30 GET, about one hour after Swigert's initial attempt at setting up PTC, the correct settings were made so that all roll thrusters were enabled. *Odyssey* and *Aquarius* settled into a near-perfect barbecue roll.

In Houston the day was nearly over and non-essential NASA personnel at the Manned Spacecraft Center had been home for several hours. In space, the crew had a welcome sleep period coming up around 01:15am Houston time on 12 April, more than 16 hours since they had been woken at the Kennedy Space Center. Now they were more than 66,000 miles (106,200km) from Earth. Launched at 13:13 hours Houston time the day before, there was nothing to suggest bad luck was stalking Apollo 13.

From space, drifting across the airwaves amid the routines of navigation platform updates, star alignment marks and systems checks were the haunting tones of the US Marines' hymn, best known from the film *The Halls of Montezuma*. Marine Corps pilot Fred Haise had the tape recorder serenading Mission Control with his tribute to a proud US military service.

ABOVE The radiator subsystem, simplified in this diagram, is designed to support two independent closed coolant loops fed with a water-glycol solution. The radiator panels are on opposing sides of the Service Module, in sectors 2 and 3, and 5 and 6. (*North American Rockwell*)

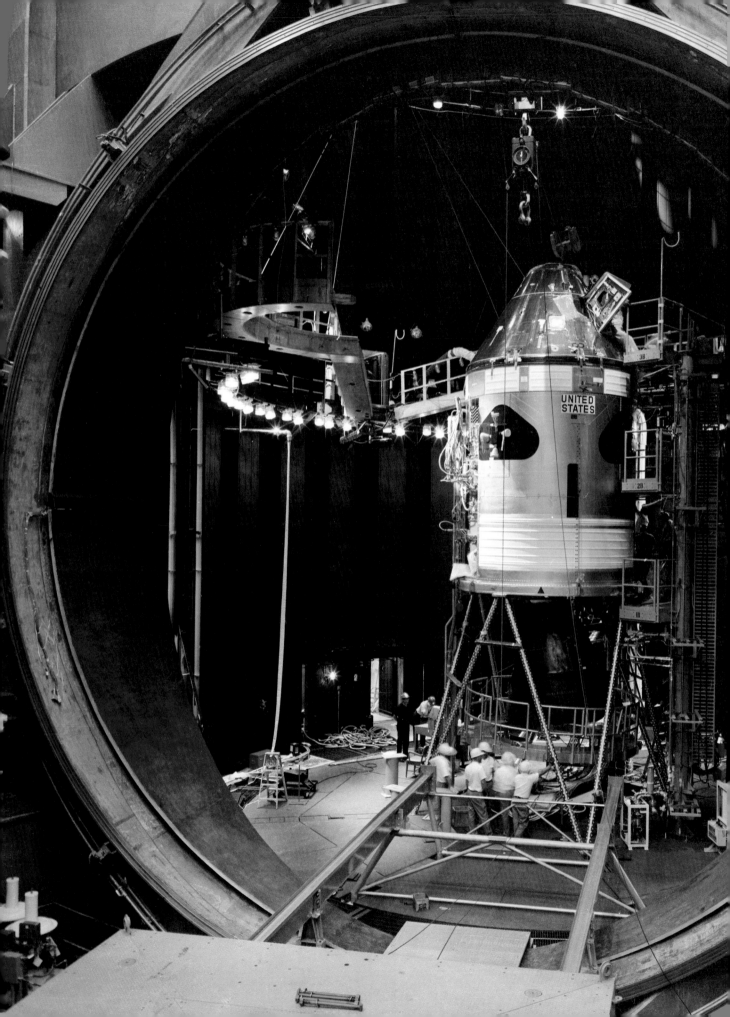

Chapter Three

Houston – we've had a problem

Public interest in Moon flights was at an all-time low. TV transmissions failed to make it on to the networks, despite efforts to stage an entertaining tour of *Aquarius*. **Apollo 13 was turning out to be just another Moon mission, until something went very wrong and alerted the crew to a potential disaster. In Mission Control, the signals were puzzling. Was it just a minor technical problem or something more sinister?**

OPPOSITE An Apollo spacecraft in the thermal vacuum chamber at the Manned Spacecraft Center, now the Johnson Space Center, Houston, Texas, located in building 32. It is 55ft (16.8m) in diameter and 90ft (27.4m) high, and is where Apollo spacecraft were exposed to a thermal and vacuum environment they would encounter during flight, simulating exposure to the cold of space and the heat of the Sun. *(NASA-JSC)*

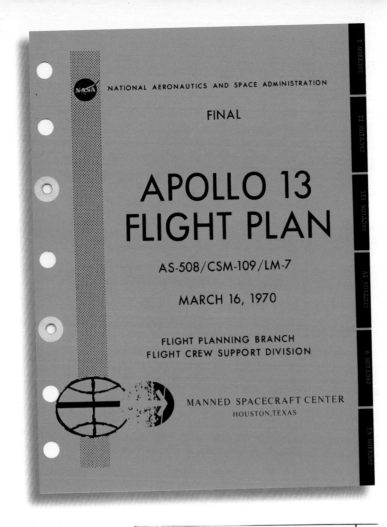

NATIONAL AERONAUTICS AND SPACE ADMINISTRATION

FINAL

APOLLO 13 FLIGHT PLAN

AS-508/CSM-109/LM-7

MARCH 16, 1970

FLIGHT PLANNING BRANCH
FLIGHT CREW SUPPORT DIVISION

MANNED SPACECRAFT CENTER
HOUSTON, TEXAS

LEFT The author's personal copy of the final edition of the *Apollo 13 Flight Plan*, which contained all the pre-planned mission events, activities which in any event would be modified as the flight progressed through a series of updates and amendments predicated by slight variations in trajectory, manoeuvre times and adjustments to the timeline. *(David Baker)*

At 3:04am on 12 April, Gene Kranz swept into the MOCR with his White Team in tow, due on shift at 13:51 GET. Capcom Jack Lousma would not have a role this shift. The flight surgeon reported that the crew had begun their sleep 20 minutes earlier and with the spacecraft in its PTC mode the shift settled into a quiet session.

Most of the functional tasks had been completed and it was time for a few 'tidy-up' actions. The electrical, environmental and consumables manager (EECOM) noted that the condenser temperature on Fuel Cell No 3 was fluctuating, but not alarmingly – it had been seen before. The mission director, Chester M. 'Chet' Lee, wanted to ask the crew some questions when they woke up about their understanding of the early CECO on the Saturn S-II stage during ascent. They were in any event scheduled for a debriefing on the performance of the Saturn V at 25hr GET.

Toward the end of the shift, the recovery officer (RECOVERY) noticed a tropical storm developing on the South Pacific. This had an impact on contingency plans for an emergency return to Earth at the T+25hr and T+35hr abort zones. This highlights the calculations made to bring the crew back to Earth fast should an emergency occur on the way to the Moon. The calculations on contingency aborts had been voiced up to the crew. Written on to a computer data entry pad usually found in the back of the flight plan, they envisaged aborts at 15hr, 25hr, 35hr, 45hr and 60hr mission elapsed time.

Abort plans were prepared based on what energy would be required to transform the circumlunar trajectory into a highly elliptical Earth orbit so that instead of going round the Moon before returning to Earth it would perform a giant loop, falling back to Earth in less time. If needed the crew would punch in the data

RIGHT Computer pads were carried in supplements to the Flight Plan, and consisted of a series of block-data panels where read-ups to the crew could be noted by an astronaut literally using pen and paper to write down the manoeuvres necessary, in this instance a P37 pad for an emergency return to Earth in the event of a serious failure. *(David Baker)*

P37 BLOCK DATA	
	GETI
X	X	ΔVT
X	X	LONG
	GET 400K
	GETI
X	X	ΔVT
X	X	LONG
	GET 400K
	GETI
X	X	ΔVT
X	X	LONG
	GET 400K
	GETI
X	X	ΔVT
X	X	LONG
	GET 400K
	GETI
X	X	ΔVT
X	X	LONG
	GET 400K
	GETI
X	X	ΔVT
X	X	LONG
	GET 400K
	GETI
X	X	ΔVT
X	X	LONG
	GET 400K

P37

to Program 37 (P37) in the guidance and navigation computer (G&NC) via a keyboard located on the main display console. A second keyboard was down in the lower equipment bay (LEB). P37 was the 'Return to Earth' abort program nobody wanted to use!

Built by the (then) Raytheon Company, the computer had very limited capacity, with 38,912 16-bit words in fixed and erasable memories, 95% of which was in the fixed memory and the balance in ferrite cores. The fixed memory was assembled from tiny nickel-iron cores woven with several thousand copper wires, each hand-wound, with each core acting as a transformer. It was indestructible and impossible to corrupt, and could not be affected by radiation or by default through human error.

Inputs to specific tied-in programs were made through a keyboard, officially the Display Keyboard (DSKY, or disskee). This had a 21-digit display and a 16-button keyboard and was used for a wide variety of guidance and navigation tasks throughout the mission. Instructions were placed in the computer via verbs (the operator) and nouns (the operand) using a coded numerical language. This simple and effective way of talking to a computer using two-digit numbers has today been replaced but it remains one of the most logical and easiest to operate.

The 'pads' upon which direct information about a specific manoeuvre calculated by Mission Control were written were sent via the keyboard to specific programs. In the case of aborts, it was Program 37. In the case of a propulsive burn, a P30, the crew member would take out his pad from the flight data file and fill in the spaces as they were read up by Capcom in Mission Control.

As an unbroken string of numbers, meaningless to the uninitiated listening to the air-to-ground, they would provide the time, delta-velocity, magnitude of the burn, burn duration and any associated information necessary for the manoeuvre. It was the recalculation of the abort pad data predicated by the approaching storm that occupied the White Team while the crew slept.

Earlier, the P37 for the T+35 contingency abort, should it be needed, would have had the crew fire the SPS engine for a ΔV of 7,733ft/sec (5,272mph or 8,483kph) at 35hr GET. This

MISSION CONTROL

There were two Mission Operations Control Rooms (MOCR) in Mission Control, both in Building 30, and MOCR-2 on the third floor was assigned to Apollo 13, as it had been to all previous Apollo flights except the first manned mission, Apollo 7 in October 1968. The room supported four rows of consoles and these could be reassigned different roles according to the programme.

For Apollo 13, the back row was assigned seating positions for senior managers including the director of the Manned Spacecraft Center, the director of Flight Crew Operations, and the Department of Defense liaison officer. Next down, the second row had positions for the public affairs officer (the only person talking to the world at large), the flight activities officer, and the flight director and assistant flight director.

The third row contained positions for the flight surgeon, EECOM (monitoring the spacecraft electrical, environmental systems and consumables), CAPCOM (the capsule communicator, the only person talking to the crew), and a series of controllers monitoring critical spacecraft systems in both the CSM and the LM, including GNC (guidance navigation and control systems), INCO (instrumentation and communications

BELOW Mission Control during the flight of Apollo 13, showing Fred Haise during the third TV transmission shortly before the accident. *(NASA-JSC)*

RIGHT Christopher Columbus Kraft (seated) was the architect of the Mission Control concept, and is seen here with Walter C. Williams (white shirt), who had been an integral part of the high-speed/high-altitude flight teams working the X-series research aircraft. *(NASA –JSC)*

ABOVE The Mission Control Center concept began at Cape Canaveral with this layout for the one-man Mercury missions of 1961–63. *(NASA-JSC)*

BELOW By the time the manned Gemini flights began in 1965, Mission Control had moved to the newly constructed Manned Spacecraft Center (now the Johnson Space Center) at Houston, Texas. *(NASA-JSC)*

systems), TELMU (telemetry, environmental and electrical systems in the Lunar Module) and CONTROL (responsible for guidance and navigation systems in the LM).

The front row included BOOSTER (monitoring the Saturn stages and consisting of engineers from the Marshall Space Flight Center at three consoles), RETRO (monitoring propulsive de-orbit burns and trans-Earth injection burns out of Moon orbit), FIDO (the Flight Dynamics Officer monitoring and calculating the flight path and the trajectory), and GUIDO (the guidance and navigation officer monitoring spacecraft computers and defining the position of the spacecraft in space).

The flight control teams for Apollo 13 had been built around four shifts, each shift serving approximately eight hours on duty – one hour working up to entering their positions in the Mission Operations Control Room (MOCR, pronounced 'moker'), six hours at the consoles, and one hour debriefing and available for press conferences etc. The four teams were denoted by colours, with the following shifts and flight directors:

Shift 1	Maroon	Milton L. Windler
Shift 2	Gold	Gerald D. Griffin
Shift 3	White	Eugene F. Kranz
Shift 4	Black	Glynn S. Lunney

Eugene F. Kranz joined NASA before the first Mercury manned flights and played a significant role in developing Mercury Control at Cape Canaveral and then Mission Control in Houston when the Manned Spacecraft Center was built there in 1962. He was appointed a flight director for the Gemini 4 mission in 1965 and was the flight director on duty when Lunar Module *Eagle* landed on the Moon.

Glynn S. Lunney planned the flight procedures for the first Mercury space flights and was instrumental in writing mission rules for all operations thereafter. He was singled out by Christopher Columbus Kraft to help develop processes for advanced manned missions from MSC in Houston and was appointed flight director for the first manned Apollo flight (Apollo 7) in October 1968.

Gerald D. Griffin joined NASA in 1960 and became a flight controller with specialities in

RIGHT Key to information display, the Eidophor screens were a great step forward in presenting trajectories and flight paths. *(NASA-JSC)*

guidance and navigation. He was the lead flight director for Apollo 12.

Milton L. Windler was brought up in rural Virginia, schooled at Virginia Polytechnic Institute and State University and had a brief career in the US Air Force before he decided to join NASA.

Gene Kranz's White Team

Flight director	E.F. Kranz
Assistant flight director	J.M. Leeper
RETRO	B.T. Spencer
FIDO	W.M. Stoval
GUIDO	W.E. Fenner
EECOM	S.A. Liebergot
GNC	B.N. Willoughby
TELMU	R.H. Heselmeyer
CONTROL	L.W. Strimple
INCO	G.B. Scott
PROCEDURES	J.R. Fucci
FAO	E.B. Pippert
SURGEON	W.R. Hawkins

The Kranz Mantra

The principles upon which Mission Control personnel dedicated their professional lives were written by Eugene F. Kranz after the Apollo fire which took the lives of three astronauts on 27 January 1967. It reads:

'From this day forward, Flight Control will be known by two words: Tough and Competent.

'Tough means we are forever accountable for what we do or what we fail to do. We will never again compromise our responsibilities...

'Competent means we will never take anything for granted...

'Mission Control will be perfect. When you leave this meeting today you will go to your office and the first thing you will do there is to write Tough and Competent on your blackboards. It will never be erased.

'Each day when you enter the room, these words will remind you of the price paid by Grissom, White, and Chaffee. These words are the price of admission to the ranks of Mission Control.'

ABOVE Seen here during the flight of Apollo 11, Mission Control matured into a modern flight-control facility that would become the template for future activity. *(NASA-JSC)*

BELOW Chains of authority and command during the flight of Apollo 13. *(David Baker)*

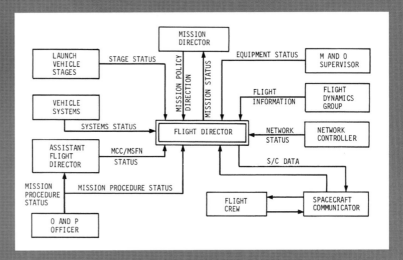

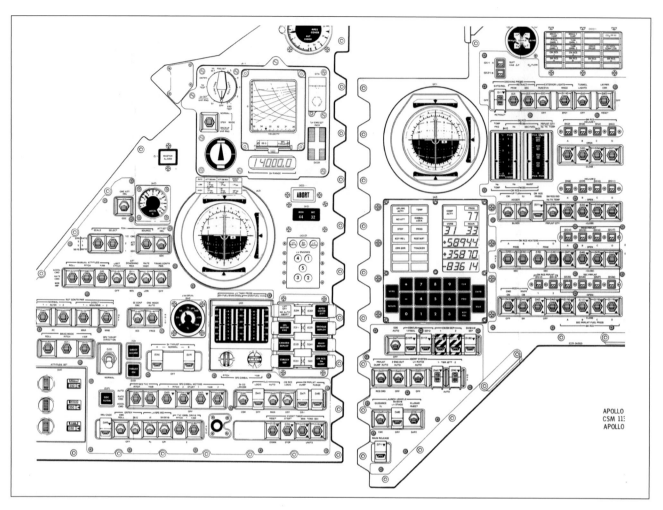

ABOVE Panel 1 and part of panel 2 for the displays aboard *Odyssey,* as viewed from the left couch. These panels display the EMS scroll (near the top of panel 1), the two FDAI 'eight balls' and, at bottom left, the 'attitude set' thumbwheels to position resolvers for each of the three axes. To its right are switches for the Service Propulsion System. Immediately below the upper-right 'eight ball' is the display and keyboard assembly for entering instructions into the Apollo Guidance Computer. *(North American Rockwell)*

BELOW The location of Service Module and Command Module reaction control system (RCS) thrusters, the latter only used after separation from the Service Module shortly before re-entry at the end of a standard mission. The SM thruster quads are capable of performing both attitude control and translation manoeuvres. *(North American Rockwell)*

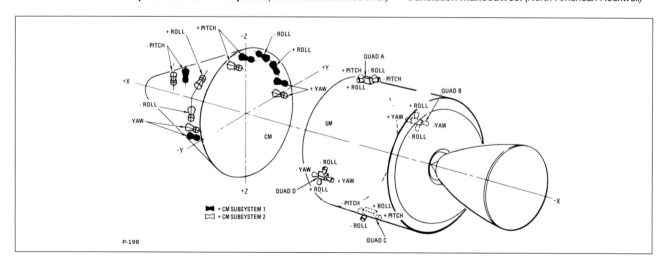

would target them for a landing in the Pacific at 165°W after entry into the atmosphere at 70:36 GET. With a storm due in that region, the longitude of landing was shifted to 155°W for entry 40 minutes earlier. To achieve that the ΔV in the T+35 abort burn would be 7,883ft/sec (5,375mph or 8,648kph).

A new course is set

Glynn Lunney's Black Team relieved Gene Kranz and his people at around 11:13am on 12 April, 22:00 GET, with Vance Brand back on the Capcom console. Just 71 minutes later the Apollo 13 crew called in, admitting to the flight surgeon that they had left their radiation dosimeters in their pressure suits, now stowed. They would voice down readings when they suited up prior to arriving at the Moon. Activity for the upcoming shift would include preparations for the all-important transfer to a non-free-return trajectory at MCC-2. FIDO (flight dynamics officer) was already conducting final burn calculations based on comprehensive tracking data. That burn would come right at the end of the Black Team's stint in the MOCR.

As for the crew, their second day brought a bout of 'space sickness', to which only about half of spacefaring astronauts are immune. During the first day Lovell noted that each crew member had red eyes and experienced a fullness in the head. Fred Haise had the distinct impression that he was hanging upside down! But after getting only around 5½ hours of sleep, their reactions to weightlessness were varied. Jack Swigert had a bad headache and took two aspirin, ate breakfast and then busied himself around the Command Module with several active tasks. This brought on a severe bout of nausea followed by vomiting so he lay down on a couch for several hours while Haise took the first of two aspirin every six hours in a successful attempt to control his own headache.

Shortly after passing the 24-hour mark, Kerwin began reading up the news to the crew as they were finishing up breakfast – all the latest sports news, plus an earthquake in the Philippines, a meeting between President Nixon and German Chancellor Willy Brandt (who had been at the launch) and an item about the latest strike by air traffic controllers. Joe reassured

LEFT Comet Bennett as viewed from Earth, an attempt to photograph it from space by the Apollo 13 crew being thwarted. *(Worth Hill Observatory, England)*

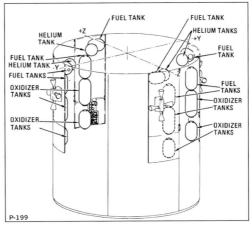

LEFT The four RCS thruster quads are attached to separate panels, each 8ft (2.44m) by 3ft (0.91m) wide, hinged on the main sector panel for access during pre-flight servicing and checks. The RCS system serves as a back-up for the Service Propulsion System for small manoeuvres and course corrections. *(North American Rockwell)*

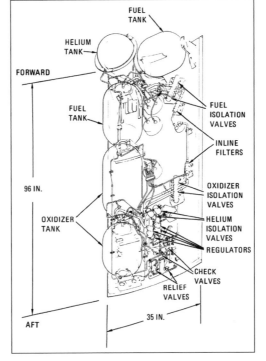

LEFT A typical SM RCS quad panel with dedicated tanks and pressurisation system. *(North American Rockwell)*

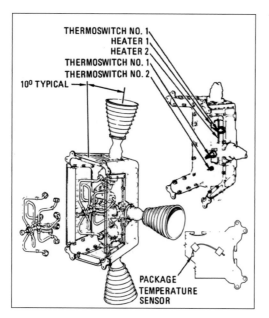

THERMOSWITCH NO. 1
HEATER 1
HEATER 2
THERMOSWITCH NO. 1
THERMOSWITCH NO. 2
10º TYPICAL

PACKAGE TEMPERATURE SENSOR

Odyssey that 'you'll be happy to know the flight controllers here in the MOCR are still on the job!'

But following an item about the last date for filing tax returns looming, Jack Swigert suddenly remembered that in the haste to get cleared for flight, he had completely forgotten to do that. And in referring to the scheduled time in biological isolation after returning from the Moon, feared that 'I may be spending time in another [type of] quarantine besides the one that they are planning for me!' And then it was back

to business, with the P37 weather avoidance update for the contingency abort at T+35hr.

The debriefing on Saturn V performance began around 25:10 GET and lasted nearly 20 minutes. Final determinations on the upcoming burn of the SPS engine were calculated and the numbers crunched for a P30 pad that Swigert would feed through the verbs and nouns of the guidance computer via the DSKY on his control panel. The all-important burn would be conducted on live TV, the camera going live at 30:13 GET. In Houston it was 7:26pm, one hour later on the Eastern Seaboard, two hours earlier in California – prime TV time; but few channels were even aware of the upcoming burn that would put Apollo 13 on a non-free-return path to the Moon.

By way of a warm-up for those watching at home while they ate their evening meal, the camera panned to window No 3 (the central side-hatch) to show, outside, the products of a 'waste water dump' – on this occasion just that (on other occasions it could be urine) – as frozen droplets spraying from a 0.055in (1.4mm) diameter dump orifice in the side of the spacecraft released liquids frozen in the vacuum of space.

With Jack Swigert in the left couch, Jim Lovell in the centre and Fred Haise with the camera on his couch to the right, TV watchers

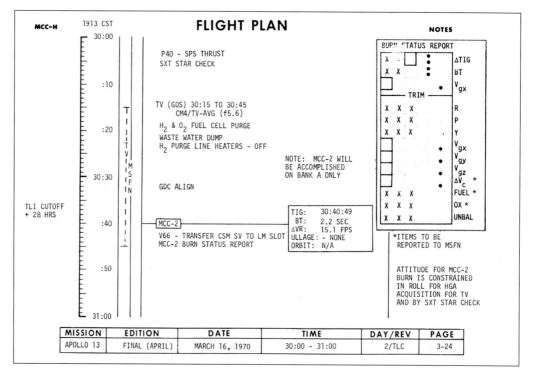

FLIGHT PLAN

MCC-H   1913 CST

P40 - SPS THRUST
SXT STAR CHECK

TV (GDS) 30:15 TO 30:45
CM4/TV-AVG (f5.6)
H₂ & O₂ FUEL CELL PURGE
WASTE WATER DUMP
H₂ PURGE LINE HEATERS - OFF

NOTE: MCC-2 WILL BE ACCOMPLISHED ON BANK A ONLY

GDC ALIGN

TLI CUTOFF + 28 HRS

MCC-2

V66 - TRANSFER CSM SV TO LM SLOT
MCC-2 BURN STATUS REPORT

TIG: 30:40:49
BT: 2.2 SEC
ΔVR: 15.1 FPS
ULLAGE: - NONE
ORBIT: N/A

NOTES

BURN STATUS REPORT

ΔTIG
BT
V_gx
TRIM
R
P
Y
V_gx
V_gy
V_gz
ΔV_c *
FUEL *
OX *
UNBAL

*ITEMS TO BE REPORTED TO MSFN

ATTITUDE FOR MCC-2 BURN IS CONSTRAINED IN ROLL FOR HGA ACQUISITION FOR TV AND BY SXT STAR CHECK

MISSION	EDITION	DATE	TIME	DAY/REV	PAGE
APOLLO 13	FINAL (APRIL)	MARCH 16, 1970	30:00 - 31:00	2/TLC	3-24

could see the control panel and watch the CMP conducting the burn. And to show the modest acceleration produced by the brief firing of the SPS engine, the crew had attached string to a pencil so that it would be seen stretching out as the vehicles moved.

Marked by ignition of the SPS engine at 30:40:49.6 GET and cut-off 3.5sec later, the MCC-2 burn placed the docked spacecraft on a trajectory that would, if uncorrected, loop around the Moon at a pericynthion (closest approach) of 69 miles (111km). When the MCC-2 burn took place the gravitational pull of the Earth had reduced the speed of the spacecraft to 3,195mph (5,140kph). As defined by tracking and computation on Earth, Apollo 13 was 139,686 miles (224,755km) from the planet. Just to show how accurate the onboard navigation system was, Haise showed TV watchers the DSKY readout of distance as computed at 139,815 miles (224,963km).

The ΔV imparted to the spacecraft by the SPS burn was 15.82mph (25.45kph), but the burn had accelerated the speed of the spacecraft by only 2.1mph (3.4kph). This was because some components of the total velocity were in changing the angle of the trajectory and not in a change in the rate at which the spacecraft were receding from Earth. On this course, Apollo 13 would

LEFT The P-30 manoeuvre pad for the SPS burn at MCC-2 indicating details of the burn, including, down the right-hand side, spacecraft weight (noun 47), pitch and yaw trim angles (noun 48), ground elapsed time for ignition (noun 33), components of delta-velocity in the three axes (noun 81), spacecraft roll, pitch and yaw attitude settings, height at apogee and perigee after the burn (noun 42), total time of the burn (ΔT), burn duration (BT), total velocity change (ΔVC), sextant shaft (SFT) and trunnion (TRN) angles. *(David Baker)*

APOLLO 13 TV SCHEDULE

DAY	DATE	CST	GET	DURATION	ACTIVITY/SUBJECT	VEH	STA
SATURDAY	APR 11	4:28 PM	03:15	1 HR 08 MIN	TRANSPOSITION & DOCKING	CSM	GDS
SUNDAY	APR 12	7:28 PM	30:15	30 MIN	SPACECRAFT INTERIOR (MCC-2)	CSM	GDS
MONDAY	APR 13	11:13 PM	58:00	30 MIN	INTERIOR & IVT TO LM	CSM	GDS
WEDNESDAY	APR 15	1:03 PM	95:50	15 MIN	FRA MAURO LANDING SITE	CSM	MAD**
THURSDAY	APR 16	1:23 AM	108:10	3 HR 52 MIN	LUNAR SURFACE (EVA-1)	LM	GDS/HSK
THURSDAY	APR 16	9:03 PM	127:50	6 HR 35 MIN	LUNAR SURFACE (EVA-2)	LM	GDS
FRIDAY	APR 17	9:36 AM	140:23	12 MIN	DOCKING	CSM	MAD
SATURDAY	APR 18	11:23 AM	166:10	40 MIN	LUNAR SURFACE	CSM	MAD*
SATURDAY	APR 18	1:13 PM	168:00	25 MIN	LUNAR SURFACE (POST TEI)	CSM	MAD*
MONDAY	APR 20	6:58 PM	221:45	15 MIN	EARTH & SPACECRAFT INTERIOR	CSM	GDS

* Recorded only
**Approval pending for satellite time

LEFT Pre-planned TV schedules from the Flight Plan, of which only the first three were actually conducted on this mission, the third being earlier due to advancing the time of entering *Aquarius*. The abbreviations under STS (tracking station) indicate Goldstone, California (GDS), Madrid, Spain (MAD), and Honeysuckle Creek, Australia (HSK). CST indicates Central Standard Time. *(David Baker)*

LEFT The 85ft antenna at Goldstone, California, a part of the global net of tracking stations nominally situated at approximate 120° intervals so that whichever way the Earth was facing there was always one deep-space station capable of communicating with the spacecraft. *(NASA)*

reach the point where the gravitational pull of the Moon became dominant at 62:49:00 GET, when it would begin to speed up. At that time it would be 219,472 miles (353,131km) from Earth and 38,921 miles (62,624km) from the Moon, moving at a mere 2,062mph (3,318kph).

Apollo 13 would reach pericynthion around the far side of the Moon at 77:28:39 GET, 28 minutes earlier than the free-return course. If no further burn was made it would loop around the Moon and return to the vicinity of the Earth, passing it no closer than 2,800 miles (4,505km). But the plan was to fire the SPS engine again, for 5min 56sec, during the close pass around the far side of the Moon, reducing the speed of the spacecraft from 5,616mph to 3,744mph, at which point it would be in lunar orbit and preparations for landing could begin.

The TV session lasted longer than planned, the crew having fun showing off their facial hair growth, Jim Lovell commenting that he was going to institute a daily shaving routine after

RIGHT The Flight Plan chart dealing with station acquisition times (AOS) and loss of signal (LOS) for the pre-planned mission, effective in displaying how the various stations were active for Apollo 13 at relative mission-elapsed times. *(David Baker)*

TABLE 1-1
S/C COVERAGE BY MSFN STATIONS USING 85-FT/210-FT DISH/ANTENNA

		GOLDSTONE (GDS)		PARKS		HONEYSUCKLE (HSK)		MADRID (MAD)	
		AOS	LOS	AOS	LOS	AOS	LOS	AOS	LOS
EARTH ORBIT				WILL NOT NORMALLY BE AVAILABLE FOR AN APOLLO 13 LAUNCH		1:00	1:06		
		1:29	1:34						
TRANSLUNAR COAST		2:51	14:16					3:59	5:42
						8:55	18:09		
		24:32	38:46					16:30	30:41
						35:17	42:52	41:05	55:34
		48:54	62:53			57:21	67:06		
		73:02	77:15					65:17	77:16
TRANSEARTH COAST								167:42	176:52
		172:25	184:21						
						178:48	190:36		
								188:59	200:48
		196:34	208:20						
						202:50	214:53		
								213:21	224:45
		221:04	232:21						
						227:01	240:46		

NOTE: AOS AND LOS TIMES SHOWN ASSUME TRACKING TO 0° ELEVATION.

the upcoming sleep session, and showing bachelor Jack Swigert 'so that all the girls will know that he's still here. Say hello to them!' But for Swigert this was no time to pose for the girls. Duty called him away to the navigation station where he was to perform some P23 star alignments and limb sightings for navigation updates to the Apollo Guidance Computer.

Meanwhile, TV viewers saw some shots of a bright Moon in the window and Jim Lovell moved the camera past Swigert to show the sleep station beneath the right couch between the underside and the bottom of the aft bulkhead. 'Vance, these hammocks by the way are very comfortable. When we first heard about them in the design of Apollo, we thought they weren't necessary, but they turned out to be a very nice device...' And with a brief 'thank you – goodbye', the TV went blank after 50 minutes 41 seconds of continuous transmission. The show was over. But it had four enthusiasts in Mission Control: Fred's wife Mary and their three children Mary M. (14), Frederick T. (11) and Stephen W. (8).

Chasing a comet

Milt Windler's Maroon Team came on shift at 31:00:00 GET, replacing Glynn Lunney, but Vance Brand remained in the MOCR as Capcom. In Houston it was 8:13pm 12 April. The new shift busied itself with preparing to take photographs of the comet Bennett, another cislunar science task. The attempt to photograph Bennett would be made at an elapsed time of 32hr using a 16mm data acquisition camera attached to the 28-power sextant. The photography would be conducted with the spacecraft as stable as possible, before the crew set up another PTC prior to the next rest period.

When the crew came to try and photograph the comet, reflections off the surface of *Aquarius* washed out the view and they were unable to get any pictures, despite an urgent effort in the MOCR involving Ken Mattingly, Vance Brand and astronaut Tony England, all in a huddle trying to find an attitude orientation that would allow one of the windows to get a visual sighting while excluding reflections. They failed. There was nothing for it but to abandon the attempt and get back in the barbecue roll.

SERVICE PROPULSION SYSTEM

The SPS was to be responsible for mid-course trajectory corrections to and from the Moon, placing the docked CSM and LM in lunar orbit, providing thrust for a back-up rendezvous with the LM ascent stage should that be necessary after lunar lift-off and for propelling the CSM out of lunar orbit and on to a trans-Earth flight path.

On 2 August 1963, North American asked NASA if Grumman's LM Descent Propulsion System (DPS) would be tested for back-up use in case the SPS engine failed before the landing, thereby providing a redundant means of getting back to Earth. That question had already been asked at a NASA management meeting the previous November, assurances being given that as a LM would be available for all Moon-bound flights the DPS would be available for such a purpose.

Apollo manager Robert Piland wanted assurance that, if called upon to save a mission, the technique would work from an engineering standpoint. On 12 March 1964 he asked North American Aviation to look into such a procedure. They conducted dynamic analysis that required modifications to the docking interface by NAA for the CSM and by Grumman for the

LEFT The Service Propulsion System (SPS) provides velocity changes for getting into and out of lunar orbit, and for mid-course corrections, as well as orbital changes around the Moon. A thermal blanket encloses the base end of the SM to prevent thermal flow up into the lower bulkhead. *(NASA-KSC)*

RIGHT The arrangement of propellant tanks and engine dictates the engineering design of the Service Module itself, a module that was to have been the lower section of the spacecraft that would have landed on the Moon prior to the decision in 1962 to build a separate Lunar Module for that purpose. *(North American Rockwell)*

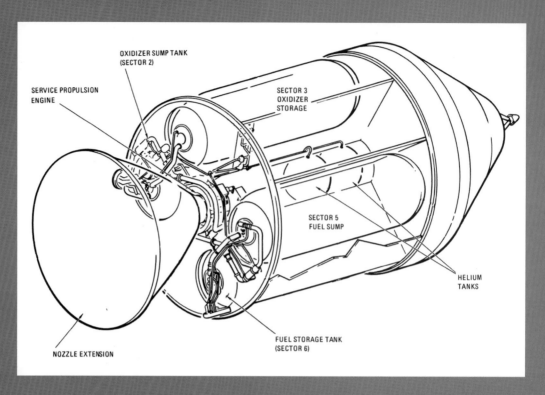

OXIDIZER SUMP TANK (SECTOR 2)

SERVICE PROPULSION ENGINE

SECTOR 3 OXIDIZER STORAGE

SECTOR 5 FUEL SUMP

HELIUM TANKS

FUEL STORAGE TANK (SECTOR 6)

NOZZLE EXTENSION

BELOW Built by the then Aerojet-General Corp, the SPS engine was one of the most fail-safe propulsion systems of its generation, and incorporated redundant design features. The engine was installed with a null offset of –2° to align the neutral thrust vector through the centre-of-mass of the CSM. *(Aerojet-General)*

BELOW A top view of the SPS engine, showing gimbal attachments and support struts together with propellant feed lines. Thrust-vector control was provided by dual servo, electromechanical actuators, capable of gimballing the engine +/–4.5° in pitch and yaw axes. Roll control was provided by the SM RCS thrusters. The gimbal actuators were contained within a sealed unit housing four electromagnetic particle clutches, two DC motors, bull jack, jack-screw and ram, bull nut, two linear position transducers and two velocity generators. A single motor and its associated pair of clutches were dedicated to independent redundant systems known as A and B. *(North American Rockwell)*

CAPACITY 800 LBS.

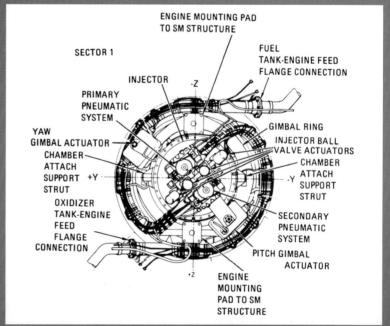

ENGINE MOUNTING PAD TO SM STRUCTURE

SECTOR 1

FUEL TANK-ENGINE FEED FLANGE CONNECTION

INJECTOR

–Z

PRIMARY PNEUMATIC SYSTEM

YAW GIMBAL ACTUATOR

GIMBAL RING

INJECTOR BALL VALVE ACTUATORS

CHAMBER ATTACH SUPPORT STRUT

+Y

–Y

CHAMBER ATTACH SUPPORT STRUT

OXIDIZER TANK-ENGINE FEED FLANGE CONNECTION

SECONDARY PNEUMATIC SYSTEM

PITCH GIMBAL ACTUATOR

+Z

ENGINE MOUNTING PAD TO SM STRUCTURE

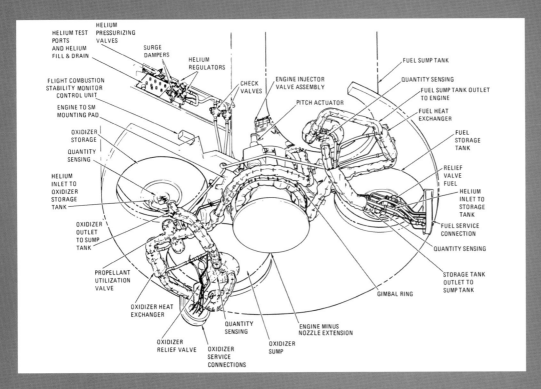

LEFT Aft view of the SPS engine components and line insulation. The propellant tanks were pressurised with helium from two 40in (101.6cm) diameter spherical tanks holding a total of 38.8ft³ (1.1m³) of gas at 3,600lb/in² (24,822kPa). The two tanks were stacked one above the other in the central tunnel running the length of the Service Module above the SPS engine. Pressurised propellants at a 1.6:1 oxidiser/fuel mixture ratio were forced into the transfer lines through the sump tank standpipe. (North American Rockwell)

LM, so that thrust from the opposite end of the docked configuration would not place undue stress on the docking ring and tunnel that connected the two spacecraft.

As built and flown the 650lb (295kg) Aerojet AJ10-137 engine had a thrust of 20,500lb (91kN), with a demonstrated life of 750sec and a sustained single firing from 0.4sec to 500sec with up to 50 restarts. In ground tests a sustained duration of 1,800sec was demonstrated, far exceeding mission requirements.

The engine had a chamber pressure of 100lb/in² (689.5kPa) with a very large expansion skirt, or nozzle, with a skirt-to-throat ratio (expansion ratio) of 62.5:1. The oxidiser was a mixture of nitrogen tetroxide (N_2O_4) and nitrous oxide, with Aerozene-50 as the fuel. Developed by Aerojet, it was a mixture of pure hydrazine and unsymmetrical dimethyl hydrazine (UDMH) in approximately equal quantities. These propellants produced a specific impulse of 314.5sec.

The engine had a height of 12ft 9in, inclusive of the 9ft 4in columbium and titanium radiation cooled expansion skirt, which had a maximum diameter of 7ft 10.5in. This was the first use of columbium, sometimes known as niobium and used as

an alloy for steels and titanium. The interior of the combustion chamber was lined with an ablative material for controlling thermal degradation, the first of its kind in such a large engine.

The hypergolic propellants were passed through concentric annuli into the injector plate, bolted to the chamber flanges with combustion stability enhanced by baffles. Any unstable combustion had to be self-rectifying within 20msec. The engine nozzle was manufactured in titanium fabricated from 16 gores and was 0.04in thick. The engine itself was contained within a drum-shaped area 44in across with a height of 48in.

This was the first engine to have a throat-gimbal design, saving more than 4ft on overall length. Thrust vector control was provided by dual servo, electromechanical actuators capable of moving the engine +/-4.5° in pitch and yaw axes. Roll control was provided by the SM RCS thrusters. The engine was installed with a null offset of -2° to align the neutral thrust vector through the centre of mass of the CSM. The gimbal actuators were contained within a sealed unit housing four electromagnetic particle clutches, two DC motors, bull jack, jack-screw and ram, bull nut, two linear position transducers and two velocity generators.

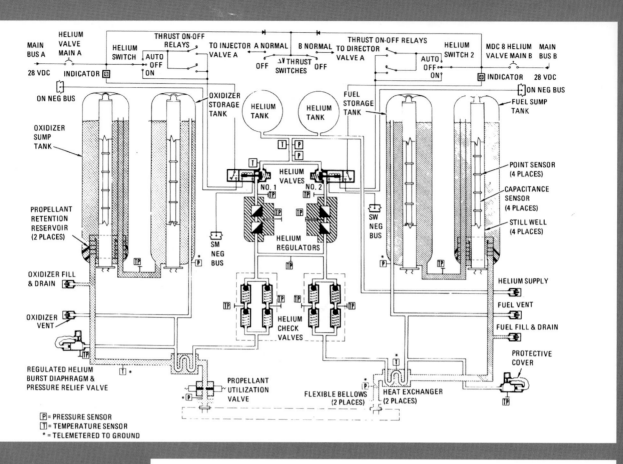

ABOVE This schematic shows the SPS propellant tank system and switch provision for controlling the flow of hypergolic propellants to the engine. Fuel and oxidiser storage tanks each had a length of 154.47in (392.3cm) with a diameter of 45in (114.3cm) and held 45% of the total load, while the sump tanks each had a length of 153.8in (390.65cm) and a diameter of 51in (129.5cm) accommodating 55% of the total. The quantities of propellant varied according to the mission. For Apollo 13 there was 15,592.7lb (7,072.8kg) of fuel and 24,937lb (11,311.4kg) of oxidiser, a total of 40,529.7lb (18,384.2kg) loaded before launch. (North American Rockwell)

RIGHT A schematic of the SPS engine-propellant feed system. (North American Rockwell)

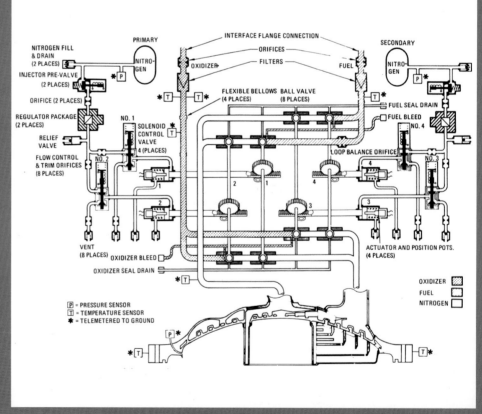

In the flight director's log the PTC was initiated at 32:15 GET, but the DAP commanded a thruster to fire which slowed down the roll rate causing divergence. It took a while to sort out the commands but PTC was finally initiated at 33:20 GET. Before the sleep period could begin there was a long list of changes to the flight plan. Before powerful computers controlling onboard systems from the ground with digital computer program updates, everything had to be entered by pen on to paper pages in the flight plan or on checklist logs.

For Mission Control, one other concern was of relatively minor importance, but the crew were warned that they would see fluctuations of 152°–160°F (67°–71°C) in the temperature at the condenser exhaust from fuel cell No 3. The other two fuel cells were relatively stable at 157°–159°F (69.5°–70.5°C), but this condition had been seen before on previous flights and Mission Control was relaxed about it. In fact, the mission was going so well that Capcom Vance Brand offered the crew an early start to their rest period, scheduled to last ten hours from 37hr GET.

The following day would see Apollo 13 closing inexorably on the Moon with an optional midcourse correction manoeuvre and a planned checkout of *Aquarius* – a period of around 90 minutes where Lovell and Haise would power up the LM and let Houston get a good long look at the telemetry. But there was concern in the MOCR about the rising pressure in the supercritical helium tank, a niggling irregularity that had haunted the mission since before lift-off. Accordingly, Vance Brand imparted to Lovell a thought that could help to ease the schedule and the minds of LM controllers:

'Jim, just an item for you and Fred to be thinking about in case you haven't been briefed on this, something that's now being talked about a little bit. To be conservative, people would like to have you read the SHe tank pressure when you go into the LM for the LM familiarisation at 58 hours. If there's no midcourse 3 [MCC-3], and it looks like there's a good chance that there will not be, why, they might want to move the LM familiarisation up from 58 to 55 hours.'

The second sleep session officially began at 37:00 GET and the crew were all tidied away

and tucked up right about on time, while in the MOCR Jack Lousma took the graveyard shift as Capcom. On board Apollo 13 the Flight Surgeon watched Jim Lovell slip effortlessly into a deep sleep. This night Lovell was the only one wearing a biomedical harness so his every heartbeat and respiration rate was coursing its way from tiny sensors attached to his body, through deep space, across the global tracking network on Earth and on to the displays in Mission Control.

Master alarms and glitches

At around 40:00 GET, 5:13am in Houston, Griffin's Gold Team relieved Windler with the crew still asleep. The RECOVERY flight controller responded to warnings that tropical storm Helen was moving swiftly into an area of the Pacific where *Odyssey* would splash down should it have to do an emergency abort at T+60 hours. So to have the manoeuvre pad ready, in the highly unlikely event that it would be required the splashdown longitude was changed from 165°W to 153°W, the latitude being 21.05°S. If there was a drastic emergency and Apollo had to bolt for home, *Odyssey* would fire its big SPS engine at 60:00:00 GET and re-enter the atmosphere 58 hours later.

For the crew, their third sleep period ended with a call to Houston at 46:42 GET, where it was 11:55am 13 April. Lovell had managed five hours' sleep, Swigert six hours and Haise nine hours. Lovell was awake early with a comment for Mission Control: 'It might be interesting that just after we went to sleep last night we had a master alarm and it really scared us...we were all over the cockpit like a wet noodle!'

In reality it had only been a low pressure reading on oxygen tank No 1 that tripped before the integral sensors saw it and switched on the heaters. By doing this a fraction of the slushy cryogenic liquid was turned to gas, raising pressure. The thought of the seasoned Jim Lovell, now on his fourth space flight, being upset by a minor problem with an oxygen tank sensor brought laughter from the MOCR.

At 47:45 GET flight plan changes for an early checkout of the supercritical helium (SHe) system in *Aquarius* were read up to the crew.

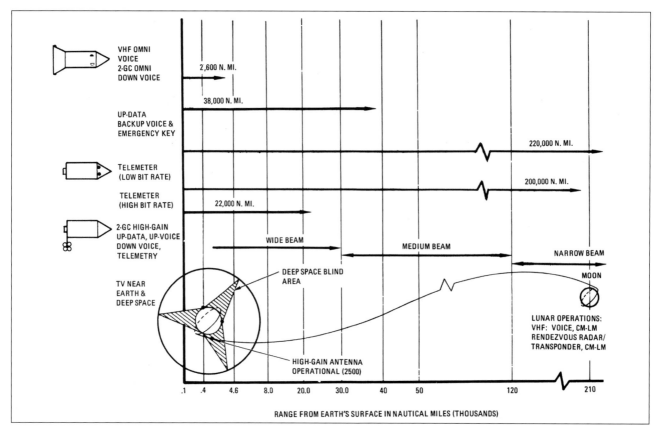

ABOVE Using nautical miles, the various VHF/AM and S-band ranges display which communication system was applicable to which phase of the mission. *(NASA)*

BELOW Attached to a 39in (99cm) boom, the High-Gain Antenna (HGA) on the base of the Service Module supported S-band communications in wide, medium or narrow-beam mode, and was a system unique to Apollo at the time of this mission. *(Motorola Inc)*

Instead of going into the LM at around 58hr GET, Lovell and Haise would advance all the checklist times by three hours, stopping PTC at 55hr for LM ingress and continuing on with activating *Aquarius*. Among the instructions voiced up to Jack Swigert was one for changes to the orientation of the High-Gain Antenna, or HGA, when *Odyssey* came out of the PTC mode for 90 minutes during LM checkout.

The HGA consisted of four circular dish antennae, each 31in (79cm) in diameter, arranged in a square pattern, 64 x 65in (162 x 165cm) in area offset 10° for optimum propagation. A central feed horn enclosure 11in (28cm) square consisted of four horns. The array was 33.8in (85.9cm) in depth and articulated for pitch and roll, carried at the end of a short boom and deployed automatically after separating from the SLA panels following TLI. In deployed configuration the antennae mast pointed down 52.25° below the Y-axis of the CSM. Four omnidirectional S-band antennae were flush-mounted to the conical side of the Command Module at 90° intervals for use in Earth orbit.

The antenna array served the Unified S-band

system (see section *The Ascent Stage*, page 176) used for all voice and data exchange with Earth from a distance of 2,880 miles (4,634km), a point at which the more conventional VHF system was out of range. Beam width for S-band went from wide (40°) to medium (4.5°–11.3°) at a distance of 34,500 miles (55,510km) to narrow beam (3.9°–4.5°) at 138,000 miles (222,000km). It became increasingly important, therefore, to control the precise pointing angle of the HGA so that its increasingly narrower beam intersected the Earth.

The instruction voiced up to Swigert was for him to halt the PTC roll at 55:00 GET and set the orientation of the HGA antennae cluster to a pitch of 23° and a yaw of 267°. In fact it should have been +5° in pitch and 237° in yaw, a difference of 35° for effective bore-sight alignment on the Earth-tracking station. Deep repetitive transients were experienced and there was a 6db decrease in uplink signal strength with a 17db decrease in the downlink strength. With the HGA in wide-beam mode both uplink and downlink should have been approximately

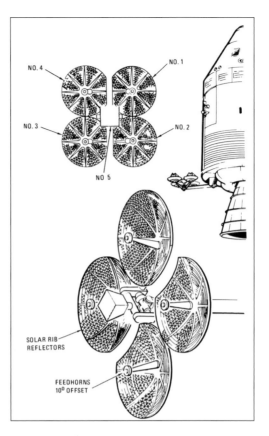

LEFT The High-Gain Antenna would be disabled by the accident to Apollo 13. A critical means of sending spacecraft data through telemetry channels and receiving uplinks and voice data, it was an essential part of the Apollo communications and data system. *(North American Rockwell)*

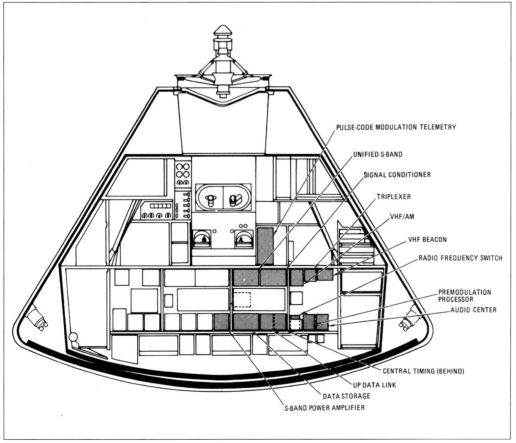

LEFT The main telecommunications equipment is located in the lower equipment bay, beneath the guidance and navigation system and its optical base, most of it contracted to the Collins Radio Co. *(North American Rockwell)*

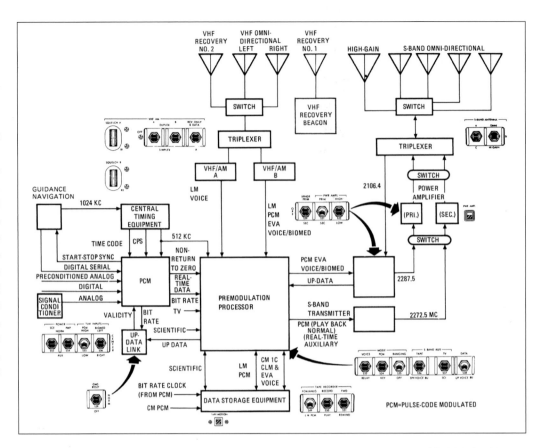

the same as the signal through the omni antennae, but they were not. The problem was solved when a reacquisition command was given ten seconds later.

It was 2:00pm in Houston (48:47 GET) when Gene Kranz and his White Team began to replace Griffin's Gold Team, his log showing he took command of the MOCR at precisely 49hr GET. The flight director had absolute control of the room and total power over decisions that could not be overruled by anyone. Several years earlier, Chris Kraft had given Kranz the job of helping to write down the duties of a flight director, and that edict was unalterable. Now Kraft was heading for the post of deputy director of the Manned Spacecraft Center and wielded a lot of power.

Initial attention was focused not only on the upcoming checkout of *Aquarius*. There had been an issue with the cryogenic hydrogen tanks. Controllers wanted to prevent a low-pressure caution-and-warning sensor (C&W) in tank No 1 triggering the master alarm during the next crew sleep cycle. The heater in tank No 2 had been switched off at 47:00 GET to see if the heater would control the pressure at

a higher level, and by 49:40 GET this had been shown to be the case. The heater had come on at a pressure of 233lb/in² (1,606kPa); the C&W limit was 224.5lb/in² (1,548kPa).

The plan was to unbalance the hydrogen quantities so that the contents of tank No 1 were 3% higher than those of tank No 2 for the duration of the sleep period. Tank No 1 heater would remain in 'auto' for the sleep period and tank No 2 heater would remain 'off' to avoid a C&W triggering a master alarm and waking the crew. But the new configuration caused the C&W tones to sound off several times while the crew were eating breakfast, cancelled with a simple flick of the switch. By the time it all settled down they would be heading toward their next sleep period, scheduled to start at 61hr GET.

Earlier, shortly after waking up at around 47hr GET, there had been an issue with oxygen tank No 2 when the gauging system suddenly failed after the crew were asked to cycle the fans in that tank. Where previously it had read a normal 82%, it now read off the scale at 100%! There was no real problem with the gauge failing. Mission Control imagined that the sensor in the tank had probably broken when the fans were activated to stir up the contents. Overall oxygen quantities could be calculated from pressure, temperature and quantity readings from tank No 1, the readings extrapolated across the two tanks.

Checking out *Aquarius*

At 50:26 GET Capcom Joe Kerwin read up to the crew a detailed set of procedures for checking out the SHe system in *Aquarius* to ensure that it would not over-pressurise before the descent to the Moon, scheduled in the flight plan for 130hr 30min. And at 51hr 10min, shortly before handing over the console to Vance Brand, Kerwin advised the crew to stir up all the cryogenic tanks and to do that more frequently than scheduled in the flight plan. Mission Control wanted a better handle on the oxygen quantities now that the sensor in tank No 2 had failed, and on the hydrogen levels, to avoid frequent alarms going off.

As events began to unfold, interest gathered momentum in the MOCR as off-duty astronaut Ken Mattingly wandered in to watch the action. Opening up *Aquarius* for the first time since launch was of little interest to an already disinterested public – they had seen it all before on previous missions – but to flight controllers and personnel it was an important event. Added to which there was the matter of the supercritical helium tank and a scheduled TV show.

TV transmissions were important for the families of the astronauts in space, and were eagerly watched; invariably a time for everyone to gather round supported by neighbours and the wives of other astronauts. They provided a valued link, albeit tenuous, with their husbands and fathers far from Earth. The business of broadcasting live TV from space had been a thorny issue with the astronauts, and the first such 'shows' had been forced on the Apollo 7 crew virtually at the last minute, and in spite of their protestations.

In different ways, TV shows from space meant different things to different people and for different reasons. But to the families of astronauts far from home they were a special connection, and there was no substitute for being in Mission Control itself as they were

broadcast. As the evening approached, Jim Lovell's wife Marilyn arrived with two of their four children, Barbara age 16 and Susan, age 11, joined by rookie astronaut Fred Haise's wife Mary. For Marilyn this was familiar territory; for Mary it was still a novelty. They would watch the show from behind the glass wall in the VIP gallery, behind and a little higher than the MOCR 'trenches'.

For an hour before starting the procedures for opening up *Aquarius*, Lovell, Swigert and Haise had what was for them their midday meal. The TV show was pegged to start at precisely 55:00 GET (8:13pm in Houston), and Lovell suggested they start on the LM early so as to complete systems checks and get a reading on the supercritical helium before it began.

The first order of business was to pump up *Aquarius* with pure oxygen to equal the pressure in *Odyssey* at 5lb/in² (34.5kPa), and that procedure began at 53:34 GET. Apollo 13 was 200,320 miles (322,315km) from Earth, travelling at a speed of 2,276mph (3,662kph). About 20 minutes later Mission Control gave the 'go' to open up *Aquarius*. Vance Brand had temporarily left the MOCR to attend a meeting on activating the LM and John Young had stood in for him at the Capcom console, the voice change momentarily confusing Fred Haise.

For this first excursion into *Aquarius*, the circular hatch at the apex of *Odyssey* was

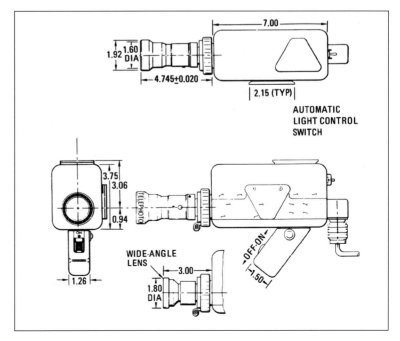

removed and stowed under one of the couches, restrained by straps so that it would not float around. This exposed the cylindrical tunnel formed by the upper section of the Command Module and the 'top' of the Lunar Module, mated by the 12 docking latches. The circular hatch into *Aquarius* contained a pressure equalisation valve to check that the two spacecraft were at the same pressure. Then a handle allowed the hatch to hinge back down inside the top of *Aquarius*. It was not removable.

Getting from *Odyssey* to *Aquarius* was achieved by floating through a short tunnel from one spacecraft to the other. While there is no up or down in weightlessness, the general configuration of *Odyssey* led to a feeling that the apex of the conical Command Module was 'up'. But on arriving in *Aquarius* everything was the other way round!

The first object encountered in *Aquarius* was the cover for the ascent engine. Twisting the body 90° around that hatbox-shaped protrusion brought the astronaut to the forward section of the Ascent Stage, where the two men would stand as they descended to the lunar surface. Turning through a further 90° would bring him to the proper 'up' position, standing on the floor of the Ascent Stage, with in front of him

RIGHT Apollo 13 telemetry equipment was provided by the Harris Corp, seen here being checked over by a technician prior to delivery. *(Harris Corp)*

two triangular windows, angled slightly forward, and a square hatch between the two positions down at floor level.

Fred Haise floated down into *Aquarius* at 54:20 GET, followed by Jim Lovell 15 minutes later. Lovell took up his station to the left, Haise to the right. Jack Swigert was in *Odyssey*. The LM batteries brought power to *Aquarius* at 54:46:15 GET. In Houston it was coming up to 6:00pm on the evening of Monday 14 April. Jack Lousma had now replaced John Young, standing in for Vance Brand. About five minutes later the crew in *Aquarius* read the supercritical helium pressure.

Mission Control estimated that if it was in the range of 660–770lb/in^2 (4,550–5,309kPa) there would be no need for further action prior to its use in pressurising the propellant tanks. If it was in the 770–800lb/in^2 (4,550–5,516kPa) range the crew would be asked to reopen *Aquarius* at 59:00 GET and check it again. If it read greater than that flight controllers were considering a short burn of the descent engine to relieve tank pressure. But that was not expected. And when they found the pressure to be 720lb/in^2 (4,964kPa), suddenly all these options were unnecessary.

Jack Lousma switched back to power from *Odyssey* at 54:58:50 GET and the crew got ready for their TV show, both men moving back into *Odyssey*, just in time to catch the evening news – if anyone outside Mission Control was watching. The TV signal was just a little late coming through, some manual antenna switching being necessary to get a solid lock-on.

'A loud bang...'

When the TV picture flickered on in Mission Control at 54:14 GET (8:27pm in Houston), Jack Swigert was on the left couch, Jim Lovell was on the centre couch playing the combined roles of cameraman and narrator, and Fred Haise had his head up in the tunnel, feigning for viewers a 'first' trip up into *Aquarius*. At the Manned Spacecraft Center only the genuinely committed newshounds were in the press room, some distance from the MOCR, watching the relay. Ever the impresario, Jim Lovell began his commentary with a scene-setter, clearly pitched to a public audience.

'What we plan to do for you today is start out in the spaceship *Odyssey* and take you through from *Odyssey* in through the tunnel into *Aquarius*,' said Lovell as Haise floated up into the tunnel once more, the Commander off the couch now and tracking Fred with the camera as he translated into the LM. Interchanging dialogues, Fred and Jim conversed for the sake of their audience as though they were just entering *Aquarius* and discovering its wonders. Nevertheless, it let viewers in on the unfamiliar feeling of floating up through the upper section of one vehicle, heads-down on to the floor of another!

Haise described how strange it felt to float into *Aquarius* 'upside down', and Lovell panned the camera to show that despite its narrow confines, weightlessness affords greater access to all corners of the spacecraft, up or down. 'One of the nice things, Jack,' said Haise, speaking to Swigert in the MOCR, 'particularly for a novice like myself, is the ease of moving around in here. It's really quite a boon to have zero-gravity as an aid.'

The TV camera showed the Moon getting ever larger. 'I can see quite distinctly some of the features with the naked eye,' said Haise, who then showed viewers the sleep positions in *Aquarius* – a hammock that could be slung between the sides of the midsection, a view made difficult by the open hatch. Lovell would sleep on a hammock, Haise having already decided he would choose the floor across the crew stations where they would stand to control the landing.

While Fred stowed the hammock, Lovell perched on the ascent engine cover and panned the camera from the interior of *Aquarius* back up through the tunnel to show Jack Swigert at the navigation station in *Odyssey*. It was 8:51pm in Houston, and, 202,000 miles (325,000km) from Earth, Jim Lovell got ready to sign off from space: 'OK Houston, for the benefit of our TV viewers we have just about completed our little inspection of *Aquarius* and now we're proceeding through the hatch gap and going back up into *Odyssey*.'

'OK Jim, it's been a great show so far.' There was one last look at the bulky probe and drogue stowed temporarily on and under Fred's couch, and a final pan across to the tape recorder from

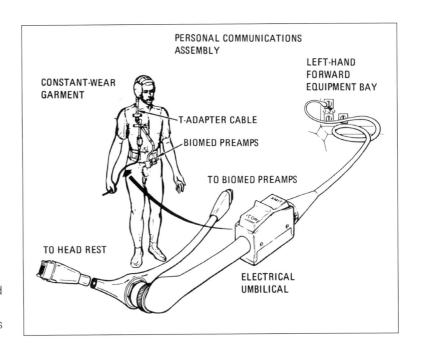

ABOVE Intercommunication and personal communication equipment incorporated a cable/headset that could be plugged in to the armrest on each couch. *(North American Rockwell)*

BELOW As data flowed down to Earth from Apollo 13, decommutators were used for unscrambling the telemetry into usable analogue information, feeding Mission Control and support rooms. Processing power was a very real problem, and there would never be enough data, limited by the capacity of the Unified S-band link. *(Harris Corp)*

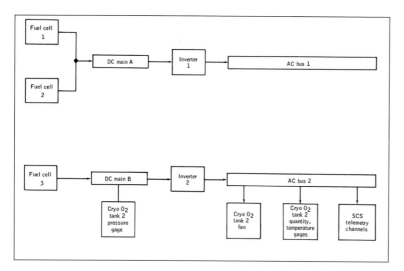

Fuel cell 1 — Fuel cell 2 — DC main A — Inverter 1 — AC bus 1

Fuel cell 3 — DC main B — Inverter 2 — AC bus 2

Cryo O₂ tank 2 pressure gage

Cryo O₂ tank 2 fan

Cryo O₂ tank 2 quantity, temperature gages

SCS telemetry channels

ABOVE The configuration of the electrical systems on the Command and Service Modules at the time of the 'big bang', showing cryogenic telemetry off main bus B. *(NASA)*

which they played music from the film *2001: A Space Odyssey*, and *Aquarius*, the opening song from the 1967 musical *Hair*. The live TV from *Odyssey* was switched off at 55:46 GET (8:59pm in Houston). Four minutes later Fred Haise was closing up the hatch into *Aquarius*.

At 55:51 GET flight controllers watched as Jack Swigert switched off the charging on battery B. There were three silver-zinc batteries in the Command Module lower equipment bay, each with 20 cells providing an open circuit of 37.2V. Sy Liebergot, the duty EECOM on the White Team, confirmed that battery B was fully charged, reading 40Ah as it should. But he wanted the cryogenic tanks stirred up one more time before setting up the PTC barbecue roll, and at 55:53 GET Jack Lousma voiced up the request: 'Thirteen, we've got one more item for you, when you get a chance. We'd like you to stir up your cryo tanks.' And then there was a suggestion from Lousma that they would shortly voice up instructions about having another go at photographing Comet Bennett.

Telemetry streaming to Earth via the Unified S-band link showed the fans to oxygen tank No 1 coming on at 55:53:18 GET. In Houston it was 9:06pm on the evening of 13 April. One second later there was a 1A spike as the pressure dropped by 8lb/in² (55.16kPa) due to destratification, as expected. One second after that, from his left couch in the Command Module, Swigert reached up and flipped the switch to power the fans in oxygen tank No 2, and a second after that pressure there too decreased by 4lb/in² (27.58kPa). As expected.

Less than two seconds after that, at

55:53:22.718, there was an electrical disturbance caused by a short in tank No 2, indicating a power transient and an almost immediate decrease of 1.2V in AC bus 2 voltage. That was *not* expected. At the same time there was an 11.1A spike in current flowing from fuel cell No 3, followed immediately by a succession of readings indicative of power coming off one fan motor, indicative in turn of an open circuit. Thirteen seconds later, at 55:53:36, the pressure in oxygen tank 2 started to rise before stabilising at 953.8lb/in² (6,576.5kPa) at 55:54.

In those 24 seconds, the stabilisation and control system showed signs of power transients, with a sudden spike of 22.9A in current flowing from fuel cell 3. And then, at 55:54:15, the pressure in tank No 2 started to rise again, followed 16 seconds later with a significant rise in temperature as well. At 55:54:43 the flow rate of oxygen to all three fuel cells began to fall, and two seconds later oxygen tank No 2 reached its maximum noted pressure of 1,008.3lb/in² (6,952.2kPa).

As the temperature in the tank began to rise sharply, converting slushy liquid oxygen into a gas, the readings went off-the-scale high, until collapsing to off-the-scale low at 55:54:52.703 GET. Just 0.060sec later the last pressure reading was showing 995.7lb/in² (6,865.3kPa), suggestive of a ruptured tank. And then, at 55:54:53.182, there was movement on all three axes as measured by the rate gyroscopes in the guidance and navigation platform, but nothing seriously violent. That lasted about a quarter of a second as pressure in oxygen tank No 1 dropped 4.2lb/in² (28.96kPa), with a 2.8A rise in electrical current from all three fuel cells.

Then something else happened. At 55:54:53.542 sudden accelerations in all three axes, but predominantly in the X (longitudinal) axis, sent a shudder through the docked spacecraft, the master alarm sounded, and the crew heard a very loud bang. There are lots of sounds in a spacecraft, noises that become familiar, such as pumps quietly whirring away, creaks and groans from materials expanding and contracting; and some, such as heater circuits working, the quiet hiss of oxygen from valves, are reassuring.

But this was different to the noises they

had heard before, sudden and unexpected. It rocked the combined spacecraft and sent a shudder through the structure. Fred Haise was up in the tunnel trying to close the hatch to *Aquarius* when he thought he saw the tunnel flexing as if one spacecraft was rocking on top of the other. In fact, Lovell's first thought was that it was Fred exercising the LM repressurisation valve.

It had been just 95 seconds since Jack Swigert switched the fans on in tank No 2 and in that time a flurry of confusion coursed through the telemetry associated with the electrical systems aboard *Odyssey*, none of which made much sense. Lovell had come back down the tunnel and floated across to the left couch. Swigert was on the centre couch.

Suddenly, just milliseconds after the bang, all telemetry from Apollo 13 died. Nothing was coming down to Mission Control. The screens went blank. The universal intake of breath was palpable – flight controllers transfixed for one brief, awful, moment. In less than two seconds data flickered back up on the monitors across the trenches in Mission Control and a slow exhalation of that same breath brought EECOM back to life. Aboard Apollo 13 the master alarm had sounded for six seconds before Swigert switched it off.

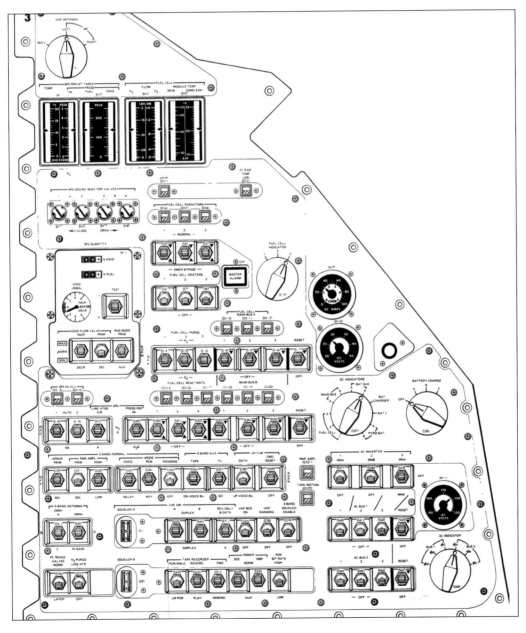

LEFT The No 3 display panel immediately in front of the Lunar Module Pilot's couch showed fuel-cell data which indicated the erratic nature of the readings, causing uncertainty in Mission Control as to the cause of the incident. *(North American Rockwell)*

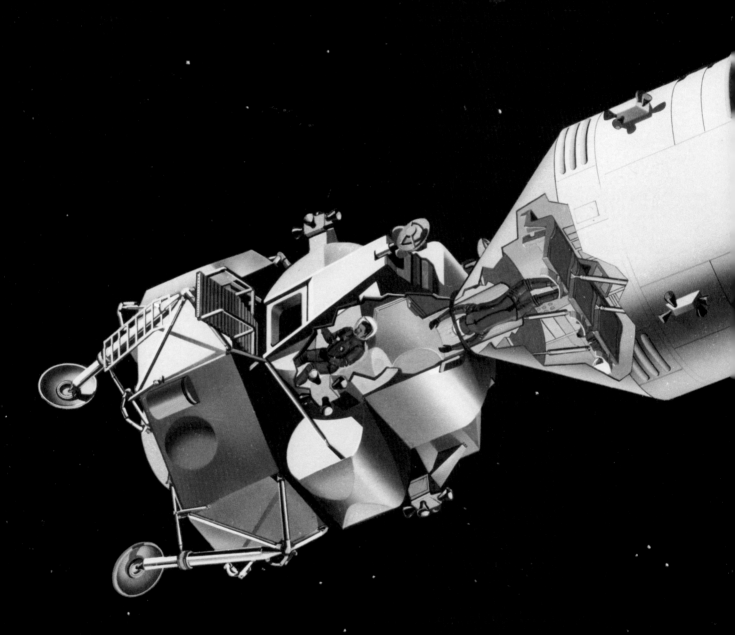

Chapter Four

Limping home

―(●)―

Reality dawned when power on *Odyssey* drained away and *Aquarius* was powered up as a lifeboat for saving the lives of the crew. First they had to fire the main engine in the Descent Stage to put them on a free-return path and then figure out how to stay alive. Soon they would be under the influence of the Moon's gravity, moving away from the Earth all the time.

OPPOSITE Within 90 minutes of the accident, Lovell and Haise were getting into the LM to power up its systems and transfer essential guidance information from *Odyssey* into *Aquarius*. *(North American Rockwell)*

While Liebergot was watching with some puzzlement the cascade of contradictory readings on his monitor, two electrical systems specialists in the mission support room switched on their voice circuits to the MOCR and shouted down the line. Larry Sheaks was the first to get to Liebergot, followed by George Bliss. Just 1.8 seconds after the first attitude excursions, as data flowed down and began to show an implausible picture of events aboard Apollo 13, Liebergot would unashamedly claim that what he saw made him almost physically sick.

The main DC bus A had dropped 0.9V to 29V, the fuel cell current was 15A higher than the value it had been prior to the brief loss of telemetry, oxygen tank No 2 was off-the-scale high (indicating failed sensors), pressure readings in that tank read off-the-scale low, suggestive of a broken feed line, pressure in oxygen tank No 1 was at 781.9lb/in² (5,391kPa) and steadily falling, and there was an unexplained temperature rise of 1.65°F in the SPS engine valve assembly. Added to all of which there was a temperature rise in a helium pressurisation bottle for one of the RCS quads on the side of the Service Module.

At 55:55:01 GET the oxygen flow rates to fuel cells 1 and 3 finally bled to zero, followed a second later by a temperature rise in the main SPS oxidiser tank in bay 3. Within seconds of that event, voltages had been restored in both bus A and bus B. But the surge of events that filled less than 30 seconds since the big bang left

no time for chatter between Lousma in Mission Control and Apollo 13. And then the silence was broken by clipped remarks from *Odyssey*.

55:55:19 Swigert: 'OK Houston…'
55:55:20 Lovell: 'I believe we've had a problem here.'
55:55:28 Lousma: 'This is Houston. Say again please.'
55:55:35 Lovell: 'Houston, we've had a problem. We've had a main B bus undervolt.'
55:55:42 Lousma: 'Roger. Main B undervolt.'
55:55:48 Lousma: 'OK, stand by 13. We're looking at it.'
55:56:10 Swigert: 'OK. Right now, Houston, the voltage is – is looking good. And we had a pretty large bang associated with the "caution and warning" there. And as I recall, main B was the one that had had an amp spike on it once before.'

As GUIDO (guidance officer), Bill Fenner had already seen a computer hardware restart and alerted Kranz. EECOM (electrical, environmental and consumables manager) confirmed a major problem with instrumentation and INCO (instrumentation and communications officer) advised that the high-gain antenna had switched from narrow beam to wide beam at 55:55:04. Faced with this and contrasting information between what the crew was seeing on their displays and what Mission Control was, or was not, seeing on their monitors, Kranz believed at first that there had been a major

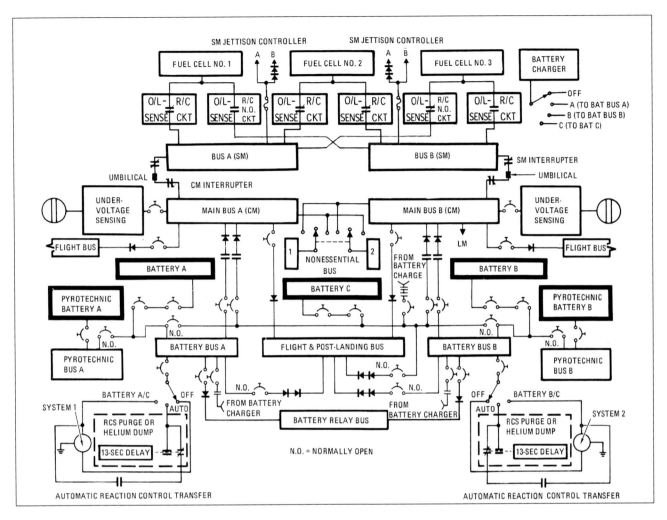

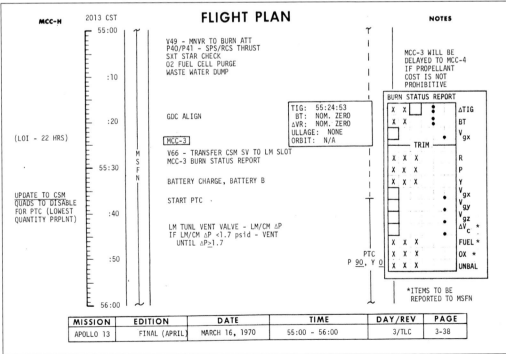

LEFT The original Flight Plan scheduled a possible mid-course manoeuvre, should one be needed, during the benign period prior to the crew getting some rest, in the hour leading up to the accident. *(David Baker)*

electrical short. He could not know how right he was in believing that; neither could he anticipate how wrong his assumption was that this had been the sole cause of the problem.

Everything seemed to point to an instrumentation failure, at worst a major electrical reset. Such was the wide diversity of contradictory readings that it was implausible to suspect a single cause for such a multiplicity of events. It certainly pointed to a major electrical failure, perhaps a massive short circuit. Before the bang, main bus B had been driving inverter 2 and AC bus 2, but when bus B went down, AC bus 2 – the circuit that provides power to the instrumentation – went with it.

Immediately after the event main bus B was off line, and at 56:00 GET the crew was advised to switch fuel cell No 1 to main A and fuel cell 3 to main B, but without any result. The talkback indicators were in the normal grey position but there was no evidence of reactants flowing to the cells. 'They are both showing zip on the flows', said Haise, as confused as Mission Control.

Bus A was holding up at 53A, which was about the most the fuel cells could provide, but voltage was levelling off at 25.5V, triggering a warning at 55:58:06 GET. EECOM asked Kranz if he should ask the crew to place fuel cell 1 on main B and No 3 on bus A, which would have the effect of easing the overload and reverse the switches so as to recycle them and perhaps bring them on line. Kranz held off on that until the controllers had a better handle on the problem – which was still confusing.

And then there was an issue with the thruster quads. GNC had been looking at the helium pressure decay in RCS quad D, and that suggested that the isolation valve had slammed shut. They correctly diagnosed the cause and did not suspect a leak. So the crew was asked to reopen the valve, which they did. At 56:07 data confirmed that the crew had reconfigured the auto-mode for the RCS thruster quads, unwanted firings threatening to waste propellant.

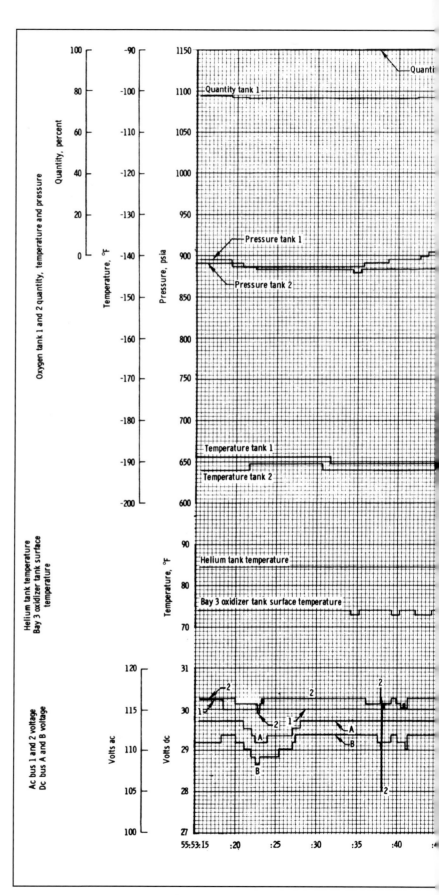

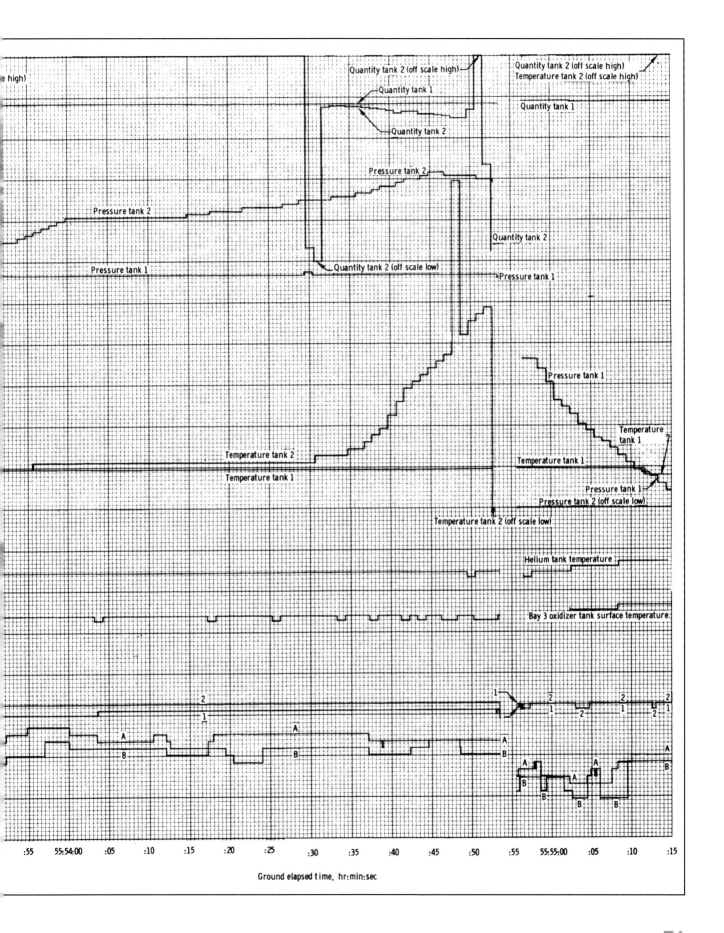

Ground elapsed time, hr:min:sec

There was still motion in the docked modules. Something was still slowly pushing them around and the inertial platform was coming close to gimbal lock, threatening to lose alignment. If that happened the spacecraft would not know what attitude it was in. Mission Control gave the crew selected switch settings to control attitude with a single thruster in each axis through main bus A. The other jets were not aligned with main bus A so as not to drag the voltage down. With heavy consumption of RCS propellant reported by GUIDO, Gene Kranz assumed the thrusters had caused a massive electrical problem that precipitated the abnormal readings.

And then, at 56:09:07 GET, Jim Lovell reported another chilling observation: 'It looks to me, looking out the hatch [window], that we are venting something. We *are* venting something out into the – into space!' As the thrusters pulsed on and off to maintain attitude, outside the double-paned circular 9in (23cm) diameter silica-glass window in the side access hatch, a steady flow of streaming particles sped past the conical Command Module. And after a few seconds for acknowledgment from Houston, Lovell confirmed: 'We are venting something... it's a gas of some sort.'

Whatever it was streaming away into space it was a valuable consumable, bleeding away from the Service Module into the vacuum of space. The central hatch window, directly above the centre couch headrest, through which Lovell had first seen it, was No 3 of five. The No 1

window, a rectangular double-paned window, was on the left side of the CM, and six minutes later when Capcom asked where it was now coming from Haise responded: 'It's coming out of window 1, right now Jack.'

The venting gas was acting like a propulsive thruster, and for the next 90 minutes Mission Control juggled incomplete data from the RCS quads, with voice reports from the crew, in an effort to understand the problem. A mixed control mode of 'auto' and 'manual' was necessary to keep the spacecraft under control. There was no significant rotation around a specific axis, just a constant attempt on the part of something to keep the spacecraft drifting off attitude. In that period, both on the ground and in space, there were much more serious conclusions to deflect attention.

While exchanges of conversation about the venting was going on, at 56:09:58 GET, fuel cell No 1 went off line, the pressure too low to maintain its operation. Just 15 minutes after the bang, that alone sealed the fate of the mission. Apollo 13 would be coming home. Even if the observed anomalies, for which there was still no certain explanation, were removed and the systems settled down, mission rules dictated that if one of the three fuel cells failed, the mission would abort. But Apollo 13 was on a non-free-return trajectory and needed a course correction to get it back on a flight path that would return it to Earth after looping around the Moon.

The uncertainty that pervaded the MOCR was not confined to the flight controllers. Sixteen minutes after the first call from Jim Lovell in *Odyssey*, the public affairs commentator went back on air at 56:11 GET to summarise for the world events that were still unclear in Mission Control and aboard *Odyssey*. The phrasing was ambiguous: 'This is Apollo Control, Houston. This rapid exchange of conversation you've heard, the main B bus is off line, fuel cells 1 and 3 are also off line, fuel cell 2 is presently on the line...' Apollo 13 was 205,582 miles (330,781km) from Earth, and moving 37 miles (60km) further away with each passing minute.

Gene Kranz called Glynn Lunney and asked him to raise senior management at the Manned Spacecraft Center and alert them to the reality that this was a very serious problem. Chains

BELOW This chart from NASA's Mission Planning & Analysis Division for the Apollo 12 mission shows the consequences of a hybrid trajectory, where the spacecraft is transferred to a non-free-return path, which Apollo 13 was now on. Unable to get back to Earth without a major engine burn, it had already lost the Service Propulsion System that could have performed that role, the only rocket motor remaining being the Descent Propulsion System in *Aquarius*. (NASA-MPAD)

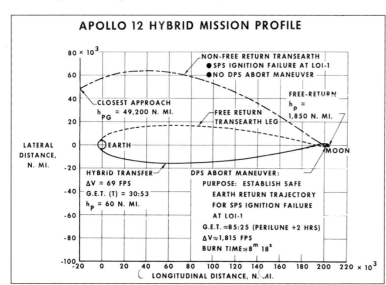

of command went right to the top. Starting that chain in motion was a serious, irreversible business and there was no going back. Soon the news would go global.

When the voice of Mission Control issued its brief summary of the situation in Houston, it was 9:24pm. Most of the news hacks had long since departed, and only a few remained behind to watch the live TV from the Moon-bound mission. Which was just as well, because none of the networks carried it live. At this time of the evening it would take about an hour to get to a downtown hotel, although many stayed out in the adjacent Clear Lake area. Most had left for bars and diners and the newsroom was all but deserted.

As 9:30pm approached, the first men to reach Houston were the 'old guard' of the Mercury and Gemini days. At the top of the management tree at the Manned Spacecraft Center was Director Robert 'Bob' Gilruth, the technical giant behind the Mercury programme, the management guru who had taken the MSC concept and built in Houston an empire that would control and operate all NASA human space flight activity. Bob Gilruth had run the Center since moving in more than eight years earlier and it was his responsibility to pass the message to NASA headquarters in Washington. As the evening wore on, that message quickly reached Richard Nixon in the White House.

Gilruth's deputy Christopher Columbus Kraft had been hauled from a shower, dripping wet, by his wife, who had answered a telephone call from Gene Kranz. When Kraft arrived at Mission Control he strode across to Kranz who, lowering his voice and turning his head to look him square in the eyes, said, 'Chris, we're in deep shit.' Not many had ever heard Gene Kranz curse. Not many would ever hear it again. Kraft, who had virtually invented the Mission Control concept, would watch as it underwent its greatest and most public test.

Also present was astronaut James A. 'Jim' McDivitt, who managed the Apollo Spacecraft Program at MSC and who had recently stood down from flight operations after commanding Apollo 9; Deke Slayton, director of flight crew operations; and Sigurd A. Sjoberg, director of flight operations. Soon Mission Control was awash with senior management. The mood

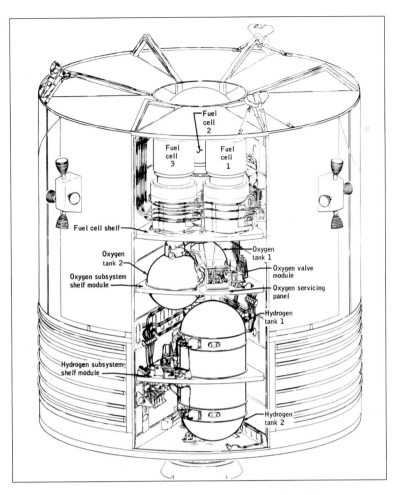

was calm, while information, discussion and clustered debate flowed constantly about the room as various members of the NASA team plugged into the events unfolding in space.

Gene Kranz, meanwhile, was weighing up the options. Venting? That could be either oxygen or hydrogen fuel cell reactants, but oxygen was essential for maintaining electrical power and for life in the Command Module. Nevertheless, Kranz polled his team for signs of any leaks. Loss of data from the quantity sensor on oxygen tank No 2 was not too problematic. After all, that could be an instrumentation failure, since AC bus 2 was down and the instrumentation circuits were on that loop; and that sensor had already been erratic.

A matter of life...

The situation was worsening – fast. At 56:15 GET, 20 minutes after the bang, Kranz asked EECOM to get his back-up people ready to stand by for using *Aquarius* as a lifeboat. At the same time EECOM asked for a major power-down in the Command Module. Electrical power was draining away and it was vital to conserve as much energy as possible to bridge the gap between powering up the Lunar Module and completely running out of electrical power altogether.

Sy Liebergot based his power-down instructions around page E1-5 of the Crew Emergency System Checklist, one of about 20 documents in the Flight Data File container. Housed in the right-hand forward stowage compartment above the right-hand couch and behind the controls and displays console, the triple-ring binder was used to take the power demand down from its present 53A to 43A. In fact the crew managed to lower demand to 41A.

While all this was going on Kranz asked the computer supervisor to pull up the Dynamic Standby Computer for working telemetry delogs, and he had set the wheels in motion for activating the 210ft (64m) Parkes, Australia, site of the Deep Space Network. It would take a day to get that up and running for mission support, but

BELOW Two separate shelves support the cryogenic fluids needed to produce electricity in the fuel cells. The two spherical titanium hydrogen tanks have an outer shell 31.9in (81cm) in height with a wall thickness of 0.033in (0.08cm), and an inner vessel with a diameter of 28.24in (71.73cm), a thickness of 0.046in (0.12cm) and a rupture pressure of 450lb/in² (3,102.7kPa). *(NASA)*

with power-down and low transmission output levels to conserve energy they would need all the receiver/transmit power they could get.

At 56:25 GET (9:38pm Houston time), Liebergot suggested putting power into AC bus 2 via inverter 1, so as to look at telemetry from oxygen tank No 2. By now, from conversations with the back room, EECOM had a good idea that two fuel cells were irretrievably lost and that this was causing the bad readings, not an instrumentation failure. EECOM also alerted Kranz to diminishing pressure in the No 1 oxygen tank. At 56:31 GET it was down to 318lb/in² (2,193kPa) and falling, and following an even deeper power-down he wanted to bring up the heaters in that tank to raise the pressure and ensure a continuing supply.

At 56:28 GET, aboard Apollo 13 the crew were in a state of information-limbo, able only to see their displays, ignorant of the swathe of conflicting data streaming down to Mission Control and unsure of their next steps. Fred Haise summed it up: 'OK Jack, the weird configuration we're sitting in right now is we have the hatch installed, we still have the probe and drogue inside the Command Module, and we're going to stay in this situation until you – kind of give us an OK to reinstall the probe and drogue...' – and after a brief acknowledgement from Lousma in Mission Control, he added '...or if necessary to use the LM consumables.'

On the ground it was fast becoming a rapid dialogue between Sy Liebergot at the EECOM console and Gene Kranz, now playing host to an increasing number of people turning up in the MOCR to watch, listen into the communications loop and give support. The mood was charged with tension but it was orderly – none of the frantic gesticulations and heated exchanges loved by the scriptwriters of TV dramas that for years after these events would forever present an over-dramatised slant on proceedings. There was work to do and a tight, disciplined approach was the only way to see it through.

At 56:34 GET Liebergot had the crew remove Command Module battery A from main bus A so as not to sap current from a power source that would be the only means of electrical energy in the Command Module for final descent through the atmosphere – if

they made it that far. Batteries B and C were not used at all prior to re-entry, and so they remained fully charged. He also had the crew isolate the surge tank, preserving vital oxygen for that final ride to a splashdown. EECOM saw the cryogenic heaters come on, drawing an extra 5A load from main bus A.

Six minutes later, based on his calculations of the rate at which the oxygen quantity in tank 1 was falling, Liebergot informed Kranz that he estimated about two hours' 'life' remaining in the Command Module. Around the room, coming up on 45 minutes since the bang sent *Odyssey*'s systems haywire, there was an increasing awareness that the sole priority was to get the crew back to Earth.

With one good fuel cell remaining, 'Let's make sure we don't do anything that's going to blow our electrical power or cause us to lose fuel cell number 2,' said Kranz, ordering the LM consoles to be manned round the clock and its personnel to work out ways of using *Aquarius* as a lifeboat, eking out its systems to support life – all the way round the Moon and back to Earth.

Prior to losing power in *Odyssey*, and because it would help to analyse the situation before everything was lost forever, at 56:48 GET Mission Control asked the crew to do a verbal report of instrument readings and displays on main display consoles 1 and 2. These were the left and centre panels of the main instrument display reaching across all three couch positions and were primarily dedicated to flight control, RCS management, cryogenic systems and environmental control instruments and switches. For the next eight minutes Jack, Jim and Fred took it in turns to give a detailed readout of what they saw on their instruments.

Meanwhile, as the LM people worked hard to get a proper set of fast-track power-up instructions to the crew for *Aquarius*, GUIDO was working with his own back-ups and experts to figure out the best way of getting home. The RECOVERY flight controller had already prepared an updated P37 contingency manoeuvre pad for a T+60 hour abort, which would require an SPS burn of 6,079ft/sec (1,853m/sec) for a return to Earth at an elapsed time of 118 hours at a position 21.05°S by 153°W. Routinely, the crew had already had that on board for some time.

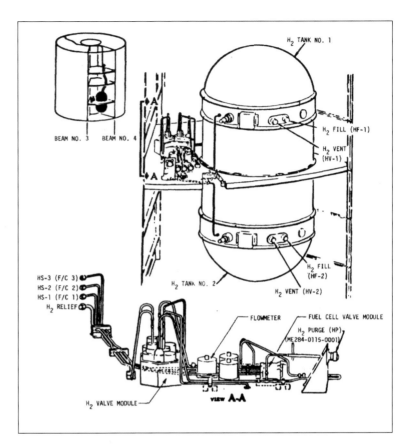

Theoretically it was within the capacity of the SPS to accelerate the return journey to a splashdown at 94 hours, with a burn of 10,395ft/sec (3,168m/sec), or 7,087mph (11,405kph), which would put the Command Module down in the Pacific at 165°W by 26.13°S. Both options would have involved transforming a loop around the Moon into a highly elliptical orbit of the Earth, apogee being reached before Apollo 13 entered the Moon's gravitational field. But the CSM was rapidly losing its ability to support any independent operations.

At 57:05 GET, 10:18pm in Houston, Gene Kranz pulled his White Team of flight controllers out of the MOCR and handed over to Glynn Lunney and the Black Team. Kranz had been at the helm of events for more than eight hours but his day's work was far from done. Whereas there were four flight control teams notionally conducting six-hour shifts, Capcoms – with three astronauts assigned to this duty – stayed for eight hours, so Jack Swigert would remain awhile as the anchor-man for the Apollo 13 crew. This was done so as not to have too many different voices changing over time.

By now, well over an hour after events had

ABOVE The two hydrogen tanks are placed one above the other, with a sleeve joining the two dewars, delivery lines carrying hydrogen up through the second shelf into the fuel cells. *(North American Aviation)*

ABOVE The centre shelf supports the two oxygen tanks side by side, the No 2 tank being outermost and adjacent to the No 1 tank inside the converging beams of sector 4. The two spherical Inconel dewars have an outer diameter of 26.55in (67.44cm) and an outer wall thickness of 0.020in (0.005cm), and an inner vessel with a diameter of 25.06in (63.65cm) and a wall thickness of 0.061in (0.15cm). Insulating fibreglass is packed between inner and outer shells. Each tank has a volume of 4.73ft³ (0.13m³) with the nominal tankage of 320lb (145kg) distributed 2:1 between fuel cells and environmental control. The top of hydrogen tank No 1 and the bottom of the fuel cells can be seen below and above, respectively. *(North American Aviation)*

pieces of data, information related to hardware, software and the general running of Apollo and the Lunar Module. Across in New York State, Grumman's LM team sprang into action. They would be in the critical path between success and failure. One employee, running red lights across the city, was stopped by a police car, only to be given a siren-wailing escort for a fast race to the Grumman plant.

Senior managers and engineers who had seen the LM through the agony years of its development were pulled in, including deputy programme manager Thomas Kelly, who happened to be in Boston on a Sloane scholarship when the call came. His wife drove him to the airport, their eldest child raised from bed to look after his five younger siblings. At Bethpage, Plant 5 was all lit up as activity at the Mission Support Center signalled a major effort already under way.

On Long Island it was nearing midnight, but employees pouring in to the Bethpage facility found floodlit car parks but increasingly fewer parking spaces. The situation was the same at countless locations across the country. On the West Coast, where many of the industrial aerospace contractors had their facilities, it was still mid-evening, but across Long Island Sound in Connecticut, Hamilton Standard had a very special role to play in getting the crew back home.

Headquartered in Windsor Locks not too far from the border with Massachusetts, the company had designed and built the LM environmental control system. That night engineers Ted Jansen and Charles Wigmore had been on duty, in case they were needed, during the checkout of *Aquarius* just before the accident. They had a speaker telephone at their desk, alongside TV monitors and cathode tube displays showing data coming down on the telemetry link. A special red telephone was a direct line to the Mission Evaluation Room in Mission Control.

NASA asked Grumman to pull all its contractors, and the Hamilton Standard programme office in Houston got a call from Bethpage. They in turn called John C. Beggs, programme manager for the life-support backpacks, who raised Andrew Hoffman, programme manager for the environmental control system, from sleep. It took him ten minutes to drive the ten miles (16km) from

taken a dramatic turn, the Manned Spacecraft Center was lighting up like a Christmas tree and Mission Control was going into battle mode. People were arriving from all over the local area, engineers turning back in to offer their services. Main contractors had field offices around uptown Houston and they in turn were connected to people at home plants where expert opinion could be on hand. Never had it been more needed.

Engineers employed at contractor plants and manufacturing facilities left their homes and drove to work, people vital for providing obscure

home to office, where he called Warren Pinter, engineering manager for the Environmental Control System (ECS), who activated a ready-list of specialist engineers at Windsor Locks. And so it went on, right across the country at countless facilities. It is estimated that 150 facilities and more than 3,500 additional engineering staff were mobilised to support the rescue of Apollo 13.

Back in the MOCR, fresh on shift, Lunney's team responded to the gradual decline in oxygen pressure, down now to $255lb/in^2$ (1,758kPa), by shutting down the supply of reactants to fuel cell 3 at 57:06 GET. The oxygen quantity continued to fall, so, suspecting a leak in fuel cell 1, the reactants to that were closed at 57:14.

At 57:24 GET, almost 90 minutes after the bang, Jack Swigert quietly reported that his displays showed the pressure in oxygen tank No 1 now down to a fraction over $200lb/in^2$ (1,379kPa) and still falling. On the ground Capcom Jack Lousma confirmed that they were watching that too, and that the temperature figures confirmed a valid reading, adding the inevitable 'It's slowly going to zero and we're starting to think about the LM [as a] lifeboat.' Liebergot did some quick sums and told Kranz it was falling at $1.7lb/in^2$ (11.7kPa) each minute.

Aboard *Odyssey* the crew too had been discussing their options. Oxygen tank No 2 was gone – its contents vented to space – but the contents of tank 1 were also bleeding away. Jack Swigert was the on-board expert in CSM systems but all three knew the basic physics. When the pressure fell to $100lb/in^2$ (689.5kPa) it would be incapable of providing a supply to the fuel cells – or to the cabin. And the temperature too was going down. As Swigert powered down the systems in the Command Module it was getting colder, down already to 58°F (14.4°C) and noticeably chilly.

Swigert offered some action: 'You want me to do a quick P52?' A P52 was an optical alignment using the Command Module sextant and telescope to update the computer on any drift in the guidance platform, a vital correction prior to any engine burn. Lousma advised the crew that the priority was to get *Aquarius* powered up and in a fit condition to sustain life. The P52 could wait.

But there was one action essential to ensuring that, if they made it all the way round the Moon and back, they could survive entry through the atmosphere when the Service Module was finally jettisoned. Like a mini-spacecraft all on its own, the CM had batteries and its own oxygen supply for supporting the crew. Stowed inside the left-hand equipment bay, the oxygen surge tank held 3.7lb (1.68kg) of oxygen at $900lb/in^2$ (6,205kPa) in a nickel-steel alloy pressure vessel 14in (35.6cm) tall and with a diameter of 13in (33cm).

The surge tank was kept topped up under normal operations by oxygen from the two cryogenic tanks. Now, with one gone and another failing, it too was running down. Liebergot's back-up guru, George Bliss, alerted him to this and a quick conference with Kranz brought a call-up to the crew from Jack Lousma: '13, Houston. We'd like you to isolate your O2 surge tank.'

At 57:36 GET, 91 minutes after the bang, the crew got another call from Lousma: 'We figure we've got about 15 minutes' worth of power left in the Command Module, so we want you to start getting over in the LM and getting some

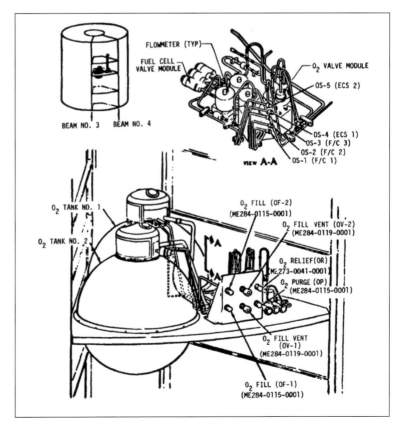

power on that.' And as if it was just a routine procedure, a minute later he asked if Swigert was ready to copy down the switch positions for putting power into *Aquarius*. That involved Fred Haise's side of the display panel, and he made the changes as they were read up from the ground. Minutes later, Lovell and Haise floated up into the LM.

There were three checklists for LM activation in the Flight Data File: the 'Activation Checklist' was the standard one for setting up the LM prior to undocking and landing on the Moon on a normal timeline; the 'Contingency Checklist' consisted of two books, the 'Two-hour Activation List' and the '30-minute Activation List'. Full contingency plans for using the LM descent engine to conduct an abort of the docked modules had been worked out and were designed into the two contingency books.

The two-hour list assumed that the main SPS had failed during the burn into lunar orbit and that the LM Descent Propulsion System (DPS) engine would be needed to fire the crew back to Earth. It included a full activation of the Primary Guidance, Navigation & Control System (PGNCS) and the inertial platform. The 30-minute list was a fast-track power-up of minimum systems for a burn without any platform activation. The problem was that the situation on Apollo 13 did not fit into any of the three checklist power-up procedures.

With power and oxygen draining away from the CSM and a need to preserve Command Module electrical power for the final dash through the atmosphere to splashdown, there was an urgent need to get the LM fully powered up to conserve CM batteries for that event. Conflicting needs were driven by the requirement to have all systems operating aboard the LM while simultaneously using as little of its power as possible to conserve those limited consumables.

There was much discussion around the consoles about how best to achieve that. Checklists were the product of extensive work in the years leading up to the flight and were not easily changed. But Glynn Lunney's team had already conducted some, albeit limited, evaluation for the previous Moon-bound flights (Apollos 10, 11 and 12) anticipating such an anomaly, so it was not entirely unknown. It was decided to stay with the standard 'Activation

Checklist' for basic detail on how to activate *Aquarius* but to use only four of its 59 pages to ensure rapid power-up.

The age of *Aquarius*

The first signals direct from the LM came into Mission Control at 57:57 GET. It was now 11:10am, just a little over two hours since a catastrophic event changed Apollo 13 from a Moon mission to a survival test. During the next hour the crew worked feverishly to stabilise the LM, power down the Command Module and configure the two modules for survival. With Jack Swigert working away in the Command Module and Jim Lovell and Fred Haise in *Aquarius*, a lengthy sequence of detailed instructions and responses flowed back and forth.

At 58:09 GET they completed an alignment of the inertial measurement units in *Aquarius* with that in *Odyssey*, handing over the coordinates of the platform from one to the other. After the platform gimbal angles were read down to Mission Control the fine-alignment torquing angles could be calculated and the Command Module Computer could be shut down, which it was one minute later.

A few minutes after that the temperature in the LM coolant loop began to rise due to heat from the newly activated systems feeding back. The crew therefore had to activate the sublimator, with the adverse effect of temporarily draining valuable water from the main tank in the Descent Stage and the two small tanks in the Ascent Stage, depleting an already marginal quantity for the return.

In Mission Control the *Odyssey* controllers were working fast to shut down their parts of the spacecraft while others manning the *Aquarius* consoles were powering up the thrusters so that the LM could control the attitude of the docked modules. At 58:32 GET, Mission Control gave Lovell and Haise computer inputs for the LM to take attitude control through its autopilot, which was completed within two minutes, but only after a particularly alert controller noticed a circuit-breaker switch had been left in the wrong position. There was no telemetry to indicate this, only the human intuition that comes with total commitment.

With confirmation that the 'hand-off' had

been effective, *Odyssey* was fully powered down, with fuel cell No 2 completely shut down at 58:38 GET and main buses A and B off two minutes later. It was 11:53pm, 1hr 45min since the bang. Never, in even the worst simulation, had complete and total shutdown of an Apollo spacecraft four-fifths of the way to the Moon been considered as a plausible situation.

But now there was another concern. After Lunney's team gave the crew the fine-torquing angles for the IMU, they asked the crew if they could see stars or recognise constellations through the forward windows in *Aquarius*, so as to conduct a platform alignment using the Alignment Optical Telescope (AOT). They could not. A mass of tiny particles – debris from whatever had happened back down in the Service Module – was floating along with the spacecraft, obscuring any pinpoint identification of distant stars.

Another issue of concern to flight controllers after LM activation was a high pressure reading in one of the two Ascent Stage oxygen tanks due to a leaky shutoff valve allowing high-pressure oxygen from the manifold to seep back into the tank. The crew were directed to use oxygen from this tank rather than the bigger Descent Stage tank. If it held up, the modest seepage would not be a problem. If it got worse and that supply was depleted, it would reduce the quantity of available oxygen to keep the crew alive.

To some extent the situation had now stabilised. When the White Team officially went off shift at 57:05 GET the situation was in crisis and very few could leave the MOCR, clinging on as their replacements slipped into the seats in front of consoles now telling a grim story. A few thought they were presiding over the last hours in the lives of three brave men; most refused to accept that. For the next hour the MOCR was full of people huddled around consoles, poring over charts, eavesdropping on small groups discussing upcoming priorities, chipping in with productive or supportive comments.

It was now a new day in Houston, and just after midnight Chris Kraft, Jim McDivitt and Sid Sjoberg sprinted over from the MOCR to the News Center, where suddenly all the journalists and reporters were clinging like leeches to every word coming down from space. TV channels

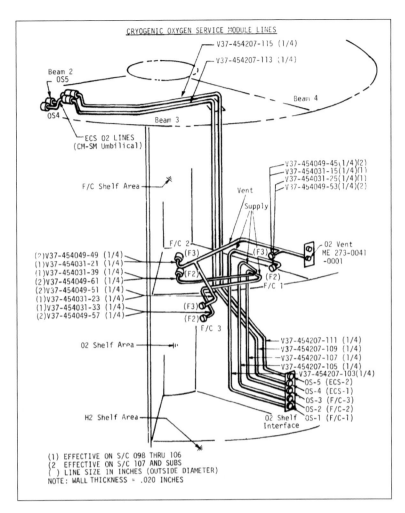

had carried 'breaking news' of events as they unfolded, and a press conference was essential to make sure that the news hacks heard it straight from the men at the centre of the action. NASA had learned long ago to be open, up-front and honest with the press. In general they had done well by the journalists and there was a mutual respect absent from many other government agencies.

The press conference began sharp at 12:20am, managed by Brian Duffy from the Public Affairs office. The crew had been on their mission 59hr 7min and it had been 3hr 12min since the accident. Kraft opened: 'Well. I guess we should start by saying that we have a serious problem [but] it appears at the present time that everything is under control... we have begun to use the LM as a device for keeping oxygen in both the Command Module and the Lunar Module.' This was measured understatement at its most cautious.

The first in a long sequence of questions

went to Mark Bloom, noted space journalist, followed by almost an hour of questions from across the floor. When the press conference ended it was getting on for 1:15am. All three managers wanted to get back to Mission Control as quickly as possible, Chris Kraft returning at a faster pace than he left, colliding with your writer as he plunged through the main doors to the MOCR.

The mood in Mission Control had changed, from a time-urgent need to get power in the LM, minimise as far as possible the amount of power being drawn from its batteries, and establish a 'lifeboat' role for Aquarius. They had bought themselves a little time, enough to think through the options without being driven by events.

The need to get back on to a free-return path was essential and, uncertain about the state of the Service Module, it could be suicide to fire up the big SPS engine. The SM was best left well alone in the hope that nothing else was brewing back down in that complex of tanks, fuel cells, pipes and rocket motors that nobody could see or get to. The DPS was the only way to get back on a free-return path, but the difficulty lay not in the magnitude of the burn – that would be relatively small – but in verifying the platform alignment when none of the crew could see stars due to the reflections from minute pieces of debris.

The use of the Descent Propulsion System to push the Apollo CSM around had been rehearsed before on the Earth-orbit flight of Apollo 9. The optimum time to fire up the engine to get back on course was 61:00 GET. But there would have to be another burn of the descent engine two hours after passing around the far side of the Moon to reduce the trans-Earth time. That burn would come almost 20 hours after the first. However, being designed for decelerating the LM down to the surface of the Moon, that engine was not designed to endure such an interval. Concerns that the ablative liner in the combustion chamber would degrade between firings was paramount, and sent specialists poring over previous test data.

For now, however, that problem could wait, because the first burn had yet to take place. There had been some debate about when to do the free-return burn, one option being to integrate it with a speed-up burn two hours after passing the far side of the moon. While negating the problem with a delayed second burn affecting the ablative liner, it would add additional requirements concerning platform alignment. While the crew were reconfiguring Aquarius, Mission Control started mapping the use of consumables: battery power, water, oxygen and the removal of exhaled carbon dioxide.

Glynn Lunney's EECOM, Clint Burton, worked up options and calculated the maximum amount of average electrical power the LM could use to keep its system running long enough to get around the Moon and back to the vicinity of Earth. All electrical power in Aquarius was provided by batteries, a total 1,600Ah in the four batteries of the Descent Stage and 592Ah in the two batteries in the Ascent Stage. At 59:00 GET, EECOM saw a total 2,113Ah remaining. Initially the LM had been using 45–50A. The selected trajectory option to get back on Earth at 142:45 GET required a maximum average use of 24A, given that for engine burns and platform alignments the systems would need much more.

While the crew used water for drinking and normally with food, it was there primarily for cooling purposes. But on a reciprocal exchange, the more electrical power used, the more water was consumed to reduce the added thermal energy and lower the overall heat load. At 59:00 GET there was about 315lb (143kg) of water remaining in the one Descent Stage tank and two Ascent Stage tanks. Usage at that time was about 6.3lb/hr (2.86kg/hr), with less than 50 hours to depletion. Water usage had to be kept below 3.5lb/hr (1.59kg/hr), corresponding to about 17A at the observed heat load. A very real problem emerged early when, with Aquarius powered up, the water consumption through the cooling system was much higher than expected.

At 59:19 GET, just 39 minutes after powering up Aquarius, EECOM saw a flow rate of 7.7lb/hr (3.49kg/hr). This was due to heat that had built up in the LM and had to be brought down, which drew heavily on the water supply. At that rate Aquarius would run dry within 40 hours! It took longer for the cabin temperature to come down than projected after initially using more water than had been expected for the power-up phase. Heat soak, so critical in

LEFT The two oxygen tanks occupy a confined space where any leaking gas will quickly build up to a high pressure in the contained area of sector 4. *(NASA)*

BELOW LEFT The three fuel cells sit atop the upper shelf, directly beneath the forward bulkhead. *(North American Rockwell)*

BELOW Tests with a full sector-4 cryogenic fuel-cell assembly are conducted on the ground prior to flight, where problems with the tanks on Apollo 13 were first discovered during pre-flight preparations. *(NASA)*

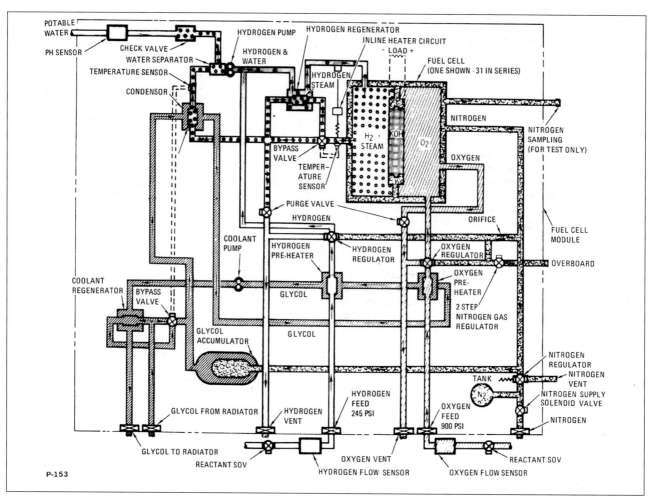

P-153

many engineering systems, is more easily acquired than rejected and can quickly lead to unintended consequences.

With 49.6lb of oxygen available in the tanks there was a hefty margin, use rate averaging 0.23lb/hr (0.1kg/hr), although a temporary rate of 0.6lb/hr (0.27kg/hr) was noted due to pressurisation of the leaky Ascent Stage tank. The amount of oxygen budgeted for Moon landing missions at the outset of LM design eight years earlier had accommodated at least two full depressurisations of the Ascent Stage for EVAs. Without any depressurisation there was a surfeit of life-giving oxygen.

Early in the crisis-management of the mission, shortly after the decision was taken to use *Aquarius* as a lifeboat, removal of exhaled carbon dioxide was regarded as the critical item. As people breathe they exhale carbon dioxide, which unless removed remains in the atmosphere and can asphyxiate if allowed to increase as a percentage of inhaled air. Lithium

hydroxide (LiOH) is an effective absorbent, and the system used in both spacecraft was to pass the oxygen across a LiOH bed and 'scrub' it as the crew went about their business.

As they were used, these LiOH beds eventually became saturated and less efficient. Under normal circumstances the maximum allowable limit of saturation was set at about 7.5mm/Hg, whereupon the exhausted LiOH canister would be swapped for a fresh one, usually every 12 hours. There were sufficient LiOH canisters stored in *Odyssey* to last a full mission of up to 300 hours. But with the CM systems powered down *Aquarius* would have to scrub the oxygen, and there was not enough LiOH to do that.

With the design expectation on the LM of about 70 man/hr divided between two men on a nominal 35hr mission independent of the CSM, *Aquarius* would now have to support 255 man/hr carbon dioxide removal for three men for the 85 hours between LM activation and separation prior to re-entry. The two primary and three secondary canisters in *Aquarius* would do that job for only 53 hours even if left to run to the higher level.

The problem was that the canisters in *Odyssey* were round while those in *Aquarius* were square, and the fittings in each spacecraft were incompatible. With *Odyssey* powered down the environmental control system in the CM was out of action and could not support CO_2 removal. For the moment, however, that problem did not require an immediate solution, so while a group went off to find one it was urgently necessary to get Apollo 13 back on a free-return course.

Clearly, consumables were going to dictate the fate of the mission. If no free-return burn was made this side of the Moon, and the spacecraft was left to coast around for a free-return correction on the way back, the Command Module would arrive back at Earth with a splashdown in the Indian Ocean at around 153:00 GET. This was unacceptable. The consumables would simply not last that long. They had to get Apollo 13 back faster than that.

What Glynn Lunney now had to decide was which option to adopt for speeding up the return journey. While obviously the thrust of the LM descent engine would not change, the amount of velocity it could impart to the docked spacecraft depended on the mass it was working against. And the repercussions of various options were exceedingly complex. In shadowing these activities your writer produced a slide rule he had expected never to use outside the classroom and finally understood why it was mandatory for all Navy pilots to carry one!

The permutations were these. After the short burn to place Apollo 13 on a free-return path, *Aquarius* had a mass of 33,452lb (15,174kg) and *Odyssey* about 62,300lb (28,259kg), of which about 50,000lb (22,680kg) was accounted for by the now inert Service Module. At this stage of the mission, the Command Module had a mass of approximately 12,300lb (5,579kg). Therefore, the now defunct Service Module alone accounted for 52% of the 95,572lb (43,451kg) mass in the docked assembly. In weightlessness there is no weight, but mass is a very present factor expressed through inertia.

In fully docked configuration the DPS in

ABOVE Fuel cells during manufacture at Pratt & Whitney, each containing 31 cells connected in series in a metal pressure jacket. *(Pratt & Whitney)*

Aquarius had a ∆V of 1,994ft/sec (608m/sec), or 1,360mph (2,188kph). If the Service Module were to be jettisoned, losing half the docked mass, the DPS could impart a ∆V of 4,830ft/sec (1,472m/sec). Some consideration was given, therefore, to jettisoning the Service Module and using the DPS to provide a long burn for a ∆V of 4,657ft/sec (1,419m/sec) to put the Command Module in the Pacific Ocean at 118:00 GET. But that would leave little or no margin for correcting flight path errors with further engine burns.

The RCS thrusters on *Aquarius* had a total translation capability of 44ft/sec (13.55m/sec). If, after the burn, a margin of error was found outside that corridor, they would be unable to make corrections for a survivable entry into the Earth's atmosphere. Trim burns, as they were called, would most likely be needed. The platform alignment was unlikely to be as precise as desired, inflicting residuals into the manoeuvre that would have to be nulled by trim burns.

RIGHT Fuel cells
undergo thermal tests
at Pratt & Whitney. The
first fuel cells used in a
spacecraft were flown
aboard Gemini 5 in
August 1965 and
continued to be used
on selected missions
for the remainder of
the programme.
(Pratt & Whitney)

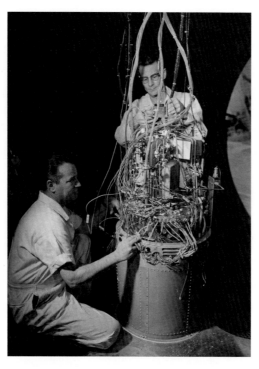

But there was another option. At 59:00 GET Mission Control had uplinked a PC+2 (two hours after pericynthion) pad to the crew for a 1,797ft/sec (577.7m/sec) DPS burn at 79:30 GET, getting the crew back at 133:15 GET in the Atlantic Ocean. This assumed no fly-by manoeuvre for free-return at 61:00 GET because at that time no firm decision had been reached about when to put Apollo 13 back on a free-return course. With a fly-by manoeuvre built in, this particular quick-return option required a ΔV of 1,997ft/sec (608.7m/sec) – far too marginal to consider.

RIGHT A technician
fits a mitten of flexible
Min-K, covered
with silicon-rubber-
impregnated glass
cloth, over the fuel-cell
bypass valve.
(Pratt & Whitney)

But the decision was made to do the free-return burn early and not combine it at PC+2, so that opened a further skein of options. Just what sort of hurry-up burn were they to do, two hours after rounding the far side of the Moon? In a meeting held in the second floor VIP room at Mission Control, the choices were examined. To get back at 118:00 GET the Service Module would have to be jettisoned and the burn performed early at PC+1, or 78:30 GET. It would need a ΔV of 4,728ft/sec (1,441.1m/sec), and with a 1° error in alignment would require a further ΔV correction of about 87ft/sec (26.5m/sec) at 105:00 GET.

Two other options were available that did not require the Service Module to be jettisoned. A 1,997ft/sec (609m/sec) burn at PC+2 (79:30 GET) would get the spacecraft back at 133:00 GET but leave a ΔV of around 50ft/sec (15.2m/sec) for nulling residuals – outside the capability for a safe margin. The next best option was to do an approximate 845ft/sec (257.6m/sec) burn at PC+2 and get back on the Pacific Ocean at around 142:00 GET. This latter option had a 3Σ probability of needing only a 22ft/sec (6.7m/sec) trim.

Assuming a PC+2, RETRO (retrofire officer) had a further speed-up possibility. By dumping the Service Module after that burn at 79:30 GET and doing another at 86:30 GET with a ΔV of 2,899ft/sec (883.6m/sec), he could get *Odyssey* on the water at 127:00 GET, but in the Indian Ocean. This velocity change was right on the limit for the DPS on *Aquarius*. Without jettisoning the Service Module he offered a fast return at 137:00 GET in the Atlantic Ocean by means of a second burn at 99:30 GET, of 1,100ft/sec (335.3m/sec). With this he could put the Command Module in the Pacific Ocean but close to the South American continent. This burn too was right on the limit of the DPS capability.

It came down to this: fast burns required jettisoning the Service Module, and having decided to do a free return correction early, the only manoeuvre to leave wide margins for residuals and correcting trajectory errors was a PC+2 for around 845ft/sec (257.6m/sec). As the feedback on consumables showed, the optimum line to tread was caution, with preserved back-up, between a super-fast return on a flawed and non-survivable flight path, or running out of life-support in *Aquarius*.

There were other considerations too. Jettisoning the Service Module would expose the base of the Command Module to an untested environment. Although the heat shield embraced the entire exterior of the conical structure, the main defence against frictional heating on re-entry was the convex base. To expose that to the extreme temperatures of space would be unwise and risky. Moreover, nobody knew what effect jettisoning the Service Module would have on the Command Module. Better to be close to home when that event was necessary, just before entering the atmosphere.

For some, the condition of everything below the Command Module was unknown, and nobody could know for certain what had happened or why. For the time being there was no attempt to find out. All resources were focused on getting Apollo 13 on the right path and balancing the time to get back with the consumables necessary to keep the crew alive. The decision as to whether to do the free-return burn early while the LM systems were powered up, or delay, power down and do it with the PC+2 burn, was made between 60:00 GET and 60:15 GET.

The choice was obvious: do it now while the systems were up rather than wait and possibly run into other problems. At least they would be on a free-return path. At 60:33 GET, Jack Lousma told the crew they were working up a data pad for the brief DPS firing to get them back on track and would a burn time of 61:00 GET, in 37 minutes, be enough time to get everything ready? Lovell thought not and asked for more time: 'Let's shoot for an hour if we can, Jack.' And with no real urgency about the precise time, it was so agreed. The burn would come at 61:30 GET. There was much to do before that could take place.

Elsewhere around the Manned Spacecraft Center, resources were being mobilised. The Apollo 14 back-up crew comprising Gene Cernan, Ron Evans and Joe Engle were busy working possible docked alignment manoeuvres in the simulator: getting to know the best way to handle the mass of the Apollo spacecraft, twice the weight of Aquarius, hanging on the end of its docking tunnel, brought unique problems. Down in Mission Control, three hours after officially going off shift, Kranz gathered up

his White Team where they were still hanging around in the MOCR and had them move to Room 210, across in a corner of the Mission Control building, leaving Lunney to get on with the free-return burn.

Room 210 was a spare data analysis room full of reference charts, strip readouts from spacecraft systems and piles of computer-generated paper containing the hard copy of numbers spewed out by the spacecraft, still moving further away from Earth. Chris Kraft had agreed, and Bob Gilruth had concurred, that Kranz should pull his team off the shift rotation, leaving Lunney, Griffin and Windler to do eight-hour stints round the clock while the White Team worked out a way to keep Aquarius working for the next 80 hours. Before long they were redesignated the Tiger Team.

What went on in that room was crucial to managing the safe return of the crew, decisions executed by the three rotating operational teams in the Mission Operations Control Room, the interface with the crew. It was not only a matter of budgeting the consumables so that they could support life aboard Aquarius and the powered-down Odyssey, but also of working through numerous calculations to balance the effect of one upon another. Switch positions had to be worked out and uplinked to the crew. It all had to be voiced up and copied down by one of the crew members, but only after all the possible mechanical and electrical repercussions had been tested in the simulators.

BELOW Eight radiators for the electrical production system alternate with eight panels on the forward skirt of the Service Module, encircling the gap between the SM and the Command Module. *(North American Rockwell)*

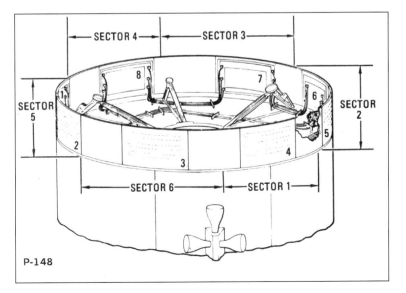

PRIMARY GUIDANCE & NAVIGATION CONTROL SYSTEM

Guidance, navigation and control of the Lunar Module was the responsibility of two independent yet cooperative systems: the primary, or PGNCS (pronounced 'pings') system and the Abort Guidance System, or AGS (pronounced 'aggs'), which effectively serves as a back-up. The PGNCS is the primary system preferably used for all mission phases but if anything goes wrong with this the AGS will take over and perform a limited function to ensure the safety of the crew during an abort. The PGNCS consists of an inertial section, an optical section and the Lunar Guidance Computer (LGC).

The inertial section established the coordinate system, or attitude frame, within which the spacecraft could operate. When aligned within an inertial reference it measured the attitude of the LM and all velocity changes, in whichever axis. Consisting of a stable platform with gyroscopes and accelerometers, the inertial measurement unit (IMU) served as a space-fixed reference aligned with stars and horizon points to establish a three-dimensional coordinate system.

This space-based orientation was known as the reference stable member matrix (REFSMMAT) and was referred to in this arcane abbreviation as the fixed coordinates periodically updated with navigational star sightings. It applied to navigation systems in both the LM and the CSM and was the only fixed and irrevocably immoveable frame within which all attitude and vector manoeuvres were planned and executed.

The IMU comprised three orthogonal gimbals, one aligned with each axis (X, Y and Z). The outer gimbal was fixed to the structure of the IMU and was aligned with the X-axis. The middle gimbal was mounted to, and perpendicular with, the outer gimbal and was aligned with the Z-axis. The inner gimbal was mounted to the middle gimbal and was aligned with the Y-axis. All three had 360° of freedom.

The three inertial gyroscopes were mounted to the stable member and were mutually perpendicular, being fluid and magnetically suspended, single-degree-of-freedom types. Three Pulsed Integrating Pendulous Accelerometers (PIPAs) were similarly suspended and provided fine measurement of acceleration in each axis. A coupling data unit converted angular information between the navigation and the guidance hardware and performed analogue-to-digital and digital-to-analogue conversions as well as processing attitude excursions.

The optical section consisted of the Alignment Optical Telescope (AOT), a periscope-type device 36in (91.4cm) long attached to the navigation base located above the astronauts' heads at their crew positions. It had a 60°

conical field of view and a moveable shaft axis. With the LM docked to the CSM only one of a possible six positions was used. For sightings, the astronaut selected a star and used the RCS thrusters to move the LM in attitude so that he could locate it within the reticle crosshair. He did this for the X and Y axes respectively, pressing a button each time to inform the Lunar Guidance Computer (LGC) of the exact time of the sighting and the gimbal angle of the inertial measurement unit.

The LGC consisted of the computer itself and a display and keyboard assembly (DSKY) and was similar in operation to the two installed in the Command Module, between the two crew member positions and above the forward hatch. In fact the LGC was virtually identical to the CMC in Odyssey. The DSKY was 8in (20cm) square with a depth of 11.2in (28.45cm) and weighed 17lb (7.7kg).

Memory in the LGC was divided into fixed and erasable sections, the total comprising 38,916 16-bit words. The fixed memory had three magnetic-rope core modules, each of which was divided into two sections. Each section had 512 magnetic cores and each core had 12 words, providing a total capacity of 36,864 words. The erasable memory comprised coincident-current, ferrite core array with a capacity for 2,048 words with a destructive readout.

Micrologic elements with resistors were diffused into single silicon wafers with one complete NOR gate (a basic logic gate) in a package the size of an aspirin tablet. Different wiring connections provide the flip-flops, registers and counters. It conducted arithmetical calculations by adding two words and moving to the next operation in 24 microseconds. Multiplication was conducted by successively adding and shifting, and division by adding complements and shifting.

Integral to the operation of the LGC, a sequence generator directed execution of programs, the central processor conducting arithmetic operations and checking information going in and out while the memory stored data and instructions.

Prioritisation was the key to sifting and filtering all the many parameters which were in a state of constant change, and time-sharing was the solution to potential overload. Like the Apollo Guidance Computer, the LGC operated in an incremental manner with one operation at a time, the highest priority item processed first. To solve guidance navigation problems, a central timer worked to Ground Elapsed Time (GET), which has also been used to denote the sequence and timing of the events in this book. The LGC had a 745.65-hour clock with a signal transmitted every second by the Manned Space Flight Network for synchronisation.

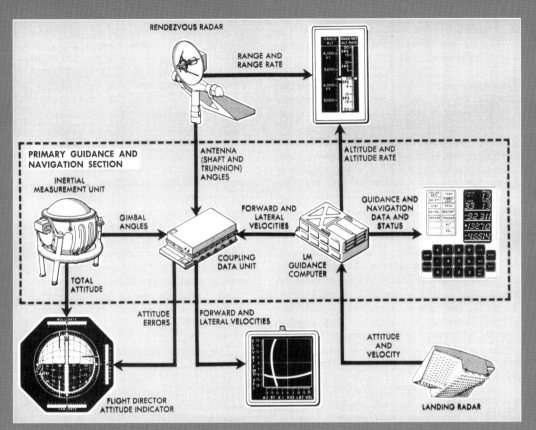

RENDEZVOUS RADAR

RANGE AND RANGE RATE

PRIMARY GUIDANCE AND NAVIGATION SECTION

INERTIAL MEASUREMENT UNIT

ANTENNA (SHAFT AND TRUNNION) ANGLES

ALTITUDE AND ALTITUDE RATE

GIMBAL ANGLES

FORWARD AND LATERAL VELOCITIES

GUIDANCE AND NAVIGATION DATA AND STATUS

COUPLING DATA UNIT

LM GUIDANCE COMPUTER

TOTAL ATTITUDE

ATTITUDE ERRORS

FORWARD AND LATERAL VELOCITIES

ATTITUDE AND VELOCITY

FLIGHT DIRECTOR ATTITUDE INDICATOR

LANDING RADAR

LEFT The PGNCS serves as an autopilot, conducting navigational duties from an established frame of reference, and includes an optical assembly and a computer sub-section for processing data and commanding engine operations. It is the brain of *Aquarius*. *(Grumman)*

BELOW The Display and Keyboard Assembly of *Aquarius* was the path into the Lunar Module's Primary Guidance, Navigation and Control System. *(Grumman)*

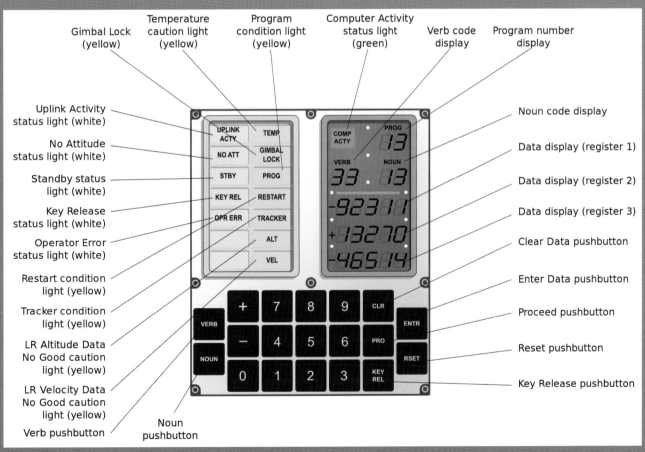

Gimbal Lock (yellow)

Temperature caution light (yellow)

Program condition light (yellow)

Computer Activity status light (green)

Verb code display

Program number display

Uplink Activity status light (white)

No Attitude status light (white)

Standby status light (white)

Key Release status light (white)

Operator Error status light (white)

Restart condition light (yellow)

Tracker condition light (yellow)

LR Altitude Data No Good caution light (yellow)

LR Velocity Data No Good caution light (yellow)

Verb pushbutton

Noun pushbutton

Noun code display

Data display (register 1)

Data display (register 2)

Data display (register 3)

Clear Data pushbutton

Enter Data pushbutton

Proceed pushbutton

Reset pushbutton

Key Release pushbutton

RIGHT The LM Environmental Control System, which would be called upon to keep the three Apollo 13 crewmembers alive for almost four days on the way back to Earth, is tested at the manufacturer Hamilton Standard. *(Hamilton Standard)*

Down in the MOCR, while all this was going on in Room 210, Glynn Lunney's people were preparing the calculations for the free-return burn. At 60:53 GET, Capcom Jack Lousma read up to Fred Haise the P40 manoeuvre pad from which he would tap in the appropriate instructions for the PGNCS to control the engine burn while Jim Lovell controlled the throttle via his hand controller.

Lovell had worried lest the attitude of *Aquarius* was incorrect, but the reflective debris outside both windows prevented visual sightings on star positions. Mission Control was not concerned. They had a good measure of the spacecraft's attitude and the burn did not call for ultra-precise alignment. John Young and Charlie Duke – the latter pulled from his sick bed – had been trying a variety of ways to give Jim Lovell a procedure for getting some sort of sighting, but to no avail.

With 30 minutes to go to the burn, Fred Haise reported that they had just deployed the landing gear from its stowed position, clearing the exhaust nozzle of the main descent engine. The manoeuvre would take place under the authority of the PGNCS system, and with the Abort Guidance System (AGS) turned off to conserve energy. The burn began at 61:29:43.5 GET and lasted 34.23sec, imparting a ΔV of 37.8ft/sec (11.52m/sec) preceded by a 7.5sec ullage burn from four RCS thrusters to settle propellant in the tanks. In Houston it was coming up on 2:43am.

Lovell held the throttle at 10% for five seconds and ramped it up to 40% for 26 seconds, although GNC data showed it at around 37%. Near enough. As a result of this burn, in approximately 16 hours Apollo 13

RIGHT This chart displays the oxygen use on a normal Lunar Module mission, where two space walks require dumping the oxygen from the Ascent Stage to the vacuum of space before the crew step out on the Moon. Fortunately for the crew of Apollo 13, the design of the LM took this into account, and without any dumping of oxygen, there was enough for the limp home. *(NASA-MPAD)*

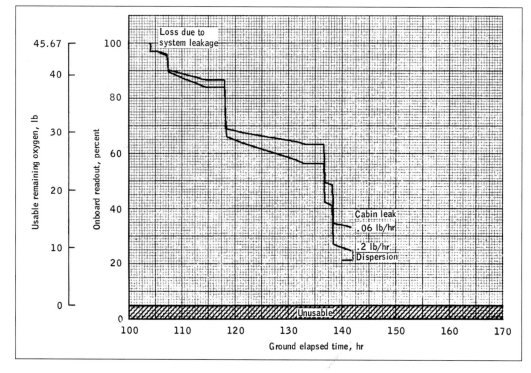

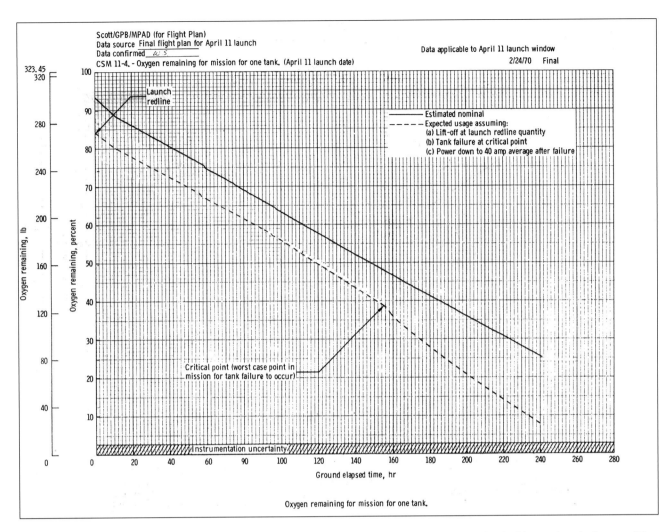

Scott/GPB/MPAD (for Flight Plan)
Data source Final flight plan for April 11 launch
Data confirmed _w S._
CSM 11-4. - Oxygen remaining for mission for one tank. (April 11 launch date)

Data applicable to April 11 launch window
2/24/70   Final

Oxygen remaining, lb

Oxygen remaining, percent

Launch redline

— Estimated nominal
— — Expected usage assuming:
(a) Lift-off at launch redline quantity
(b) Tank failure at critical point
(c) Power down to 40 amp average after failure

Critical point (worst case point in mission for tank failure to occur)

Instrumentation uncertainty

Ground elapsed time, hr

Oxygen remaining for mission for one tank.

would pass across the Moon 158.32 miles (254.74km) above the surface instead of the 74.65 miles (120.11km) previously imparted by the SPS engine when it shifted to a non-free-return path almost 31 hours before. *Aquarius* had used 408.6lb (185.34kg) of propellant in the brief burn, its mass now down to 33,452lb (15,174kg).

As a result of the free-return burn, Apollo 13 would reach Earth at 152.5 hours elapsed time, in the Indian Ocean south of Mauritius; but the PC+2 burn was being worked up to cut that by about ten hours, reducing the remaining mission duration to 81 hours – still more than three days to go. The first priority now was to power down as far as possible and then establish some reasonable form of passive thermal control.

Around 3:15am on 14 April (62 hours into the mission), things began to look just a little better, the remaining consumables eking out beyond the time now being talked about for

ABOVE In planning Apollo 13, consideration had been given to the possible consequences of an oxygen-tank failure, and this chart, carried along in the Flight Plan, shows the worst time at which such an event could occur. At around 155 hours ground elapsed time, Apollo 13 should have been in lunar orbit and would have to immediately head home, a trade-off being made with the amount of oxygen required for the fuel cells and that needed for cabin atmosphere. *(NASA-MPAD)*

the fast return. Budgeting of electrical power against requirements also had to accommodate a need to put power back into Command Module battery A, which had been used to compensate for the loss of power from the Service Module prior to powering up *Aquarius*. It was also felt desirable to have the crew put heat back in to *Odyssey*'s guidance system. But the decision was made to ride it out rather than use more LM power at this early stage in the recovery plan.

Officially, power-down began at 62:44 GET,

with usage dropping to a little over 26A. PTC would be conducted through the Digital Auto-Pilot (DAP). Apollo 13 reached the point where the gravitational pull of the Moon became dominant five minutes later, from where it began to speed up. At that time it was 219,472 miles (353,131km) from the Earth and 38,921 miles (62,624km) from the Moon, moving at a mere 2,062mph (3,318kph), but inexorably towards its nearest neighbour in space.

The crew had now been awake for more than 16 hours and there was little they could do until rounding the Moon in 15 hours' time, so a basic sleep plan was worked out leaving one man awake. Lousma told the crew they recommended Haise getting his head down at 63:00 GET for six hours, while Lovell and Swigert got a bite to eat and remained on watch. After they all had a meal at 70:00 Lovell

ABOVE Astronauts and flight controllers gather, pensive in mood, as the critical burn nears to put Apollo 13 back on a free-return path home. *(NASA)*

CENTRE With control now solely managed by *Aquarius*, the sidestick hand controller attached to one of the arm rests in the LM uses the same axial alignment as the two hand controllers in *Odyssey* for controlling the attitude of the docked spacecraft. The geometry of the four thruster quads on the Ascent Stage made controlling the two vehicles difficult. *(Grumman)*

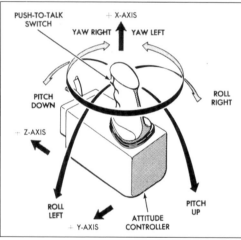

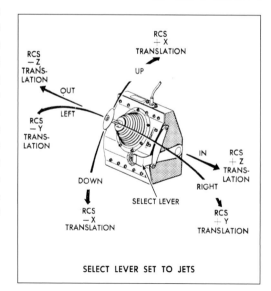

BELOW LEFT The Thrust/Translation Hand Controller (TTHC) is used for manoeuvres either firing the RCS thrusters to change velocity or flight path, or to control the Descent Propulsion System. It has a lever on the right side for selecting 'Jets' or 'Throttle'. In the 'Jets' position the TTHC commands thrusting through the RCS quads. *(Grumman)*

BELOW The TTHC in the 'Throttle' setting provides control authority over the Descent Propulsion System and RCS firings as shown in this diagram. *(Grumman)*

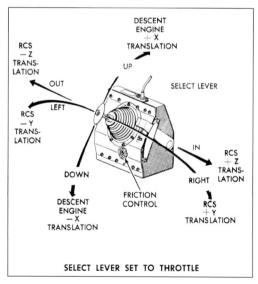

and Swigert would get six hours' sleep before all would be awake for going round the Moon and preparing for the all-important PC+2 burn.

Sleep, a euphemism under such circumstances, would be taken down in *Odyssey* away from the area where the temperature was already falling to a distinctly uncomfortable low. While there was no oxygen flowing from *Odyssey*, the supply from *Aquarius* kept the pressure up in the darkened Command Module. Only light from *Aquarius* reflected down off the cold, steel walls of *Odyssey*. The noises from *Aquarius* were a distraction, pumps whirring and gurgling, thumps from the RCS thrusters as they periodically fired to keep Apollo 13 in its PTC mode.

Nothing the crew had on board to wear was conducive to keeping warm, their thin flight coveralls being made of fireproof material incapable of retaining warmth. But personal comfort was not a consideration to any but the medics watching in the blind without telemetry to show the true condition of the astronauts, now beginning to feel the effects of a very long day. Try as he might, Fred Haise could get little rest, his sleeping bag as useless as the coveralls at keeping him warm, his body heat impossible to retain.

Communications had been poor for some hours after initially powering up *Aquarius*, but as one by one the separate antennae and communications equipment were switched off it got much worse. At 63:05 GET Mission Control passed up a preliminary P40 manoeuvre pad for a PC+2 burn of 890ft/sec (271.3m/sec), getting the crew back on Earth in the mid-Pacific at 142:40 GET. EECOM saw an average 27A load on electrical power with use rates settling down to expected power-down levels.

After the burn at 79:30 GET *Aquarius* would go into deep power-down for at most only life-support and communication functions. Based on an average load of 25.6A, the batteries would last until 142 hours, the water until 138:30 GET, and the oxygen until 233 hours, long after the crew was back home. Calculations from Gene Kranz's Tiger Team and a host of experts and specialists across the country suggested that a maximum 15.4A could be allowed to make the consumables last.

By 63:50 GET the PTC mode run by the

DAP was erratic and difficult, so Lunney's team decided to go to a manual yaw command of 90° every hour, saving further electrical power and thruster propellant. In addition the power amplifier for the communication system was taken off line, saving a further 2.5A but increasing background noise on the communication circuit. Also, various schemes for getting a P52 platform alignment were being looked at, either by using the AOT in *Aquarius* as Apollo 13 passed into the shadow of the Moon, or by using the optics in *Odyssey* but on minimum power for all.

With consumables looking bleak at worst and only just manageable at best, the Black

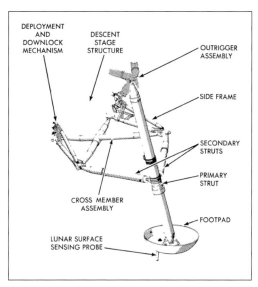

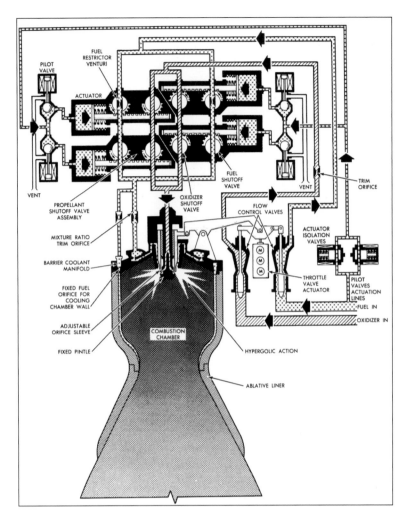

ABOVE The Lunar Module Descent Engine was arguably one of the most challenging developments of Apollo. Space Technology Laboratories (later renamed TRW) received the contract from NASA in May 1963, after Grumman had selected Rocketdyne to build this engine. Nobody had built a throttleable engine with this performance in the early 1960s, and the technology was cutting edge. TRW built the engines for the LM and the Rocketdyne work was cancelled. *(TRW)*

ABOVE An operating schematic of the throttleable Lunar Module Descent Engine. *(Grumman)*

RIGHT Operation of the Descent Engine could be conducted through automated or manual control, or with a hybrid of each, as displayed in this schematic. *(Grumman)*

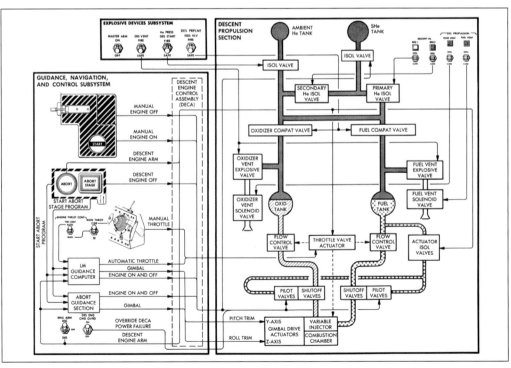

Team discussed once again the option of ditching the Service Module and doing a superfast burn to get the crew back by 118 hours. But the word from Room 210 – which had by now become a dispersal centre for a lot of other rooms in a lot of places across the country where contractors and subcontractors were burning the night oil – was to keep it on and hold to the PC+2 burn. Weather was predicted to be good at the recovery site northeast of New Zealand, with winds at 15 knots and 4ft (1.2m) seas.

After a mere two hours, Fred Haise could suffer the chill and the transient sounds no longer and drifted back down into *Aquarius*. Houston had already expressed concern at the build-up of carbon dioxide in *Odyssey*, where there were no fans to circulate the oxygen drifting across from *Aquarius*: 'What we're recommending is that you take the Commander's hose in the LM and put a cap over the red return hose and then figure out a way to fasten the hose so they blow up in the CSM by extending them up as far as possible. And we'll get some flow off the blue side circulating up and around the Command Module and to keep the CO_2 level down,' said Jack Lousma at the Capcom console.

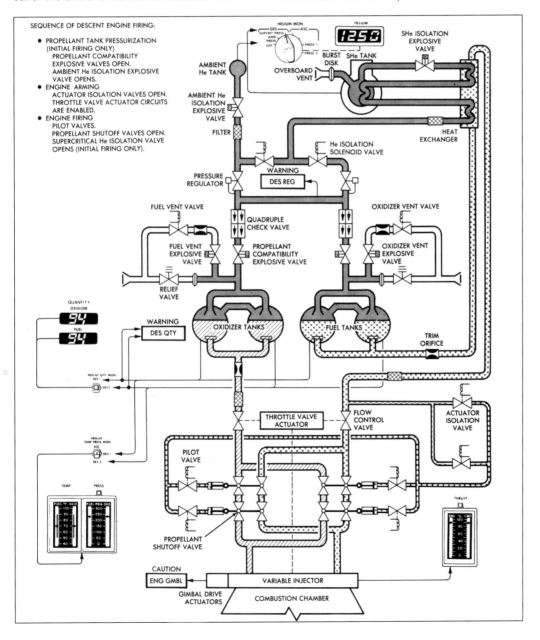

LEFT The descent-engine flow-control diagram for the Lunar Module Descent Engine (LMDE).
(Grumman)

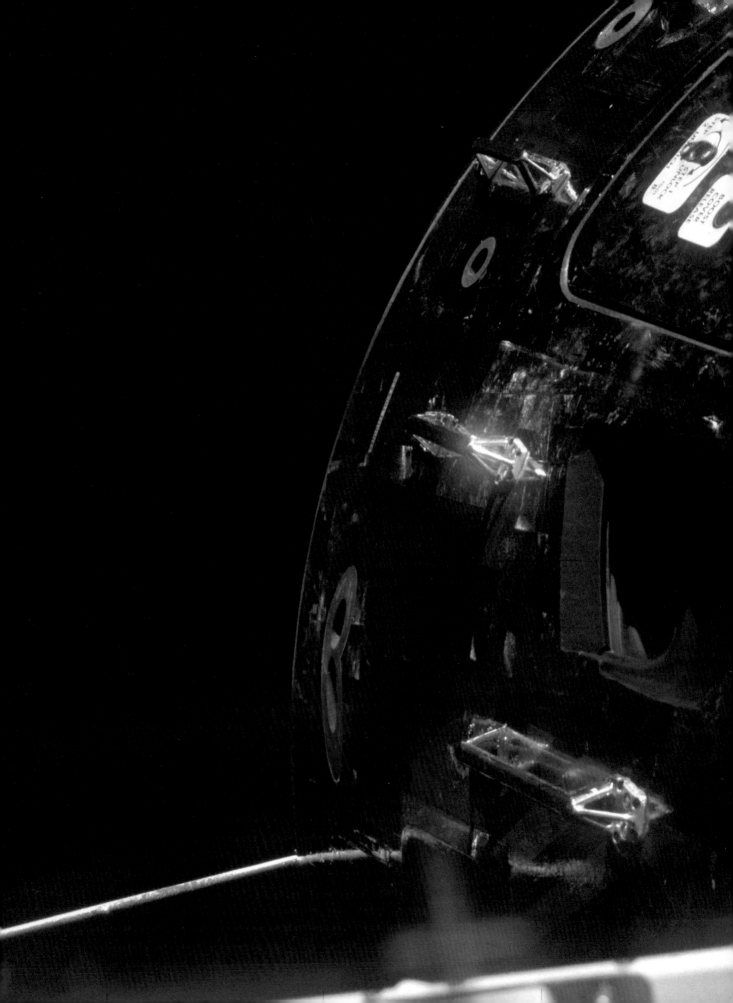

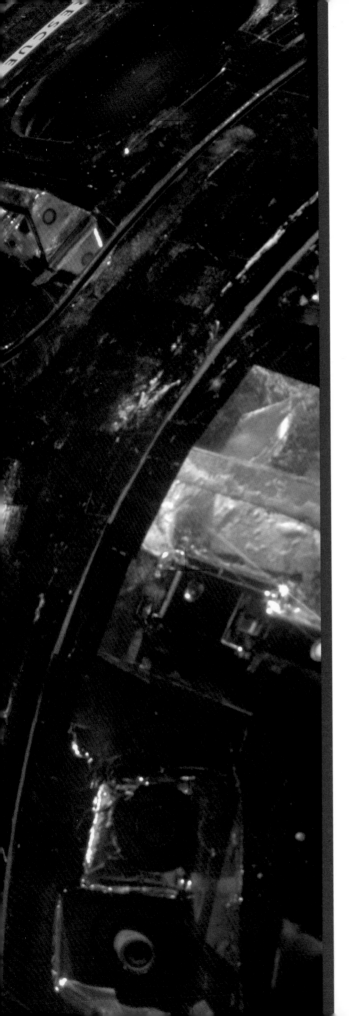

Chapter Five

The big hurry-up

Rounding the Moon, Apollo 13 was now on course back to Earth. But not fast enough to reach home before electrical power and water ran out. And then there was the problem of exhaled carbon dioxide poisoning the astronauts with their own breath. What to do? Frantic efforts in Houston and across the United States mobilised teams of scientists, engineers and technicians to figure out solutions. But time was running out.

OPPOSITE The view out the Lunar Module window looking across the conical Command Module, now all but lifeless save for the air pumped across from *Aquarius*. (NASA)

As America was waking up on 14 April the news stations were buzzing with dramatic reports broadcast across the country. At 6:40am (65:27 GET for Apollo 13) NASA Administrator Thomas O. Paine touched down at Ellington Air Force Base and was rushed to the Manned Spacecraft Center with a police escort. Less than an hour later, alerted by NASA staffers in Washington DC, President Nixon called Paine for an update, demanding to be kept informed. Less than a year earlier, unbeknown to the public, he had filed away a prepared statement to read out on nationwide TV should Apollo 11 have failed and the crew been stranded on the Moon. Now, he wondered, would he have to read a different but equally tragic announcement?

From around the world offers of help came flooding in, naval forces placed at the disposal of the American government should recovery be needed from some remote location. Religious leaders called for prayer, the Pope spoke to 10,000 people in the Basilica invoking a hope for the crew's survival, and in Australia national TV showed *Lost in Space* with captions updating viewers on Apollo 13's situation.

In space, the crew was getting by as best they could. Jim Lovell was trying to catnap while dealing with intermittent instructions from the ground. It was not yet his 'official' sleep period. Officially too, Jerry Griffin's Gold Team would be signing on at 67:00 GET, but more than an hour early they were in the MOCR,

coming up to speed on the free-return burn, the consumables status and the continuing, and annoyingly frustrating, problems with the communications circuits.

Before completely replacing Lunney's people, the new team would confer with the outgoing shift about the procedures for trying to get an alignment for the platform, and that was causing some concern. The MOCR was filling up fast, and for a couple of hours there were two complete shifts conferring with their opposite numbers, coming up to speed on procedures and getting feedback from astronaut Eugene Cernan, working the LM simulator trying to find a way to get star alignments though a debris-contaminated sky.

Director of flight crew operations Deke Slayton was there, with Apollo mission director Chester M. 'Chet' Lee (by whom the writer was tasked with going away to get some numbers crunched and with whom he would work on Shuttle manifests 15 years later), with Apollo programme director Rocco Petrone, Bob Gilruth and Chris Kraft, and several more who found it impossible to go home. Most telling was that Milt Windler and most of his Maroon Team were there also. They had been in the MOCR for some time. In fact, although approaching 8:15am, almost no one had managed to sleep.

The formal shift change took place at 67:00 GET. In Houston it was 8:13am. Having worked a double shift, Jack Lousma finally handed the Capcom chair to Joe Kerwin, who would now pick up conversation with the crew. On an increasingly scratchy link, the crew switched to vox (voice actuated) mode, in which the downlink was activated whenever a crew member spoke. This was always a dangerous move. Casual conversation could so easily be misunderstood.

At 68:03 GET, Lovell and Swigert were taking stock and trying to work out appropriate antenna configurations and switch positions for optimum air-to-ground when the Commander turned to his LM pilot and said, almost laconically, 'Well, I'm afraid this is going to be the last lunar mission for a long time.' In the garbled sounds coming down from *Aquarius*, Fred's response was lost.

Ever the diplomat, Jim Lovell never really explained exactly what he meant by that, but

BELOW Home for three-and-a-half days, the cramped confines of the Ascent Stage looking across the Commander's window in this crew cabin without couches. *(NASA)*

the press picked up on it and refused to let go. Was it a cynical statement that the Nixon administration would use this as an excuse to shut down the programme; was it a swipe at NASA; or a resignation that whatever had caused the accident, engineers would be a long time finding the reason and correcting it for the remaining flights to the Moon?

The scratchy communications were made worse because *Aquarius* was transmitting on the same frequency as the transponder on the spent Saturn V third stage, the S-IVB that had powered them on their way to the Moon almost three days earlier. That problem would go away after the S-IVB stage hit the Moon as planned to provide shock waves for the Apollo 12 seismometer to measure. The LM and the S-IVB were never meant to operate together at the same time, and with the power amplifier in *Aquarius* switched off the signals were fighting each other.

Then, at 68:45 GET, Kerwin asked Lovell to make provision for when they would have to start using carbon dioxide scrubbers from *Odyssey* to keep down CO_2 levels. 'We just thought of something; namely, that we probably should get the lithium hydroxide canisters out of the Command Module reasonably soon just to make certain that they don't stay in there and possibly swell up until they'd be hard to get out. I wouldn't wake up Fred for that, but it's something you should do possibly before you go to sleep.' And turning to Swigert he sent the CMP off up into *Odyssey* to do just that.

In Mission Control, Glynn Lunney finally left the MOCR more than an hour after the official shift handover, and walked across to Building 1 and the main auditorium for a press conference, taking with him his RETRO Tom Weichel, EECOM Clint Burton, LM G&N (guidance and navigation officer) Hal Loden and TELMU (LM telemetry, environmental and electrical systems officer) Merlin Merritt. With them on the podium would be Major General David 'Davy' Jones, the US Air Force commander of recovery forces.

The press conference started at 9:35am in a packed room – the biggest available for public affairs activity, usually set aside for major announcements. Preceded only by the one hurriedly convened conference in the early hours of the morning, it was the first time the

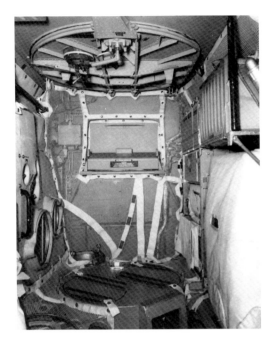

LEFT Looking aft from the forward crew area, across the covered Ascent Engine to the rear of the mid-section in *Aquarius*, to the area where, on a normal mission, the crew would place their Personal Life Support System (PLSS) backpacks. Note the closed hatch to *Odyssey*, which on Apollo 13 remained open throughout the period between power-up and close-out on the way home. *(NASA)*

world's press got to hear a full description of events over the past 12 hours.

While the news hounds waited for word about the final decision on whether to cut loose the Service Module and do a superfast return, in space the astronauts were settling into survival mode. The first three hours after the accident had been a desperate scramble to shut down *Odyssey* and power up *Aquarius* to buy time to make survival decisions. Since then the spacecraft had been put back on a trajectory that would bring them back into the close vicinity of Earth, working out a way to make the consumables last. Now it was a waiting game

BELOW Communications through the LM systems was a vital link for voice and data transfer, and this signal processor was one important part of that equipment. *(Collins)*

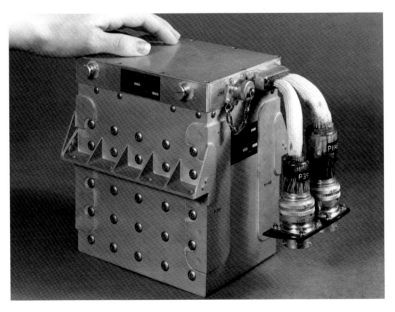

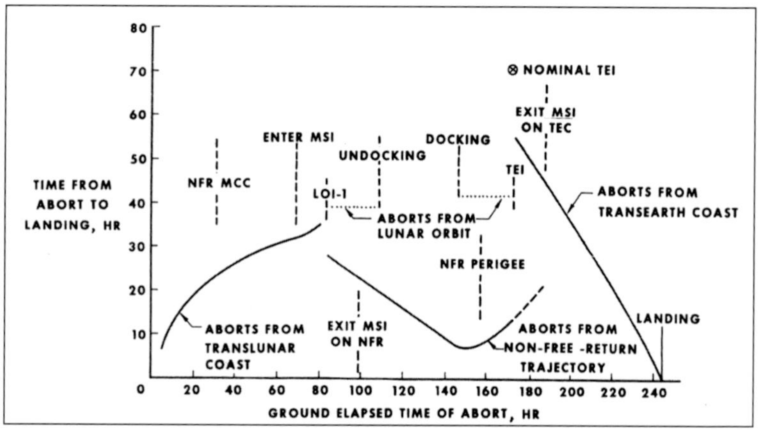

ABOVE Abort plans were extensively prepared before each mission, and for lunar flights were standard, worked out on simulators and on slide rules. This chart displays the return time on the vertical axis and the mission elapsed time along the horizontal axis. MSI infers the Moon's Sphere of Influence, or equigravisphere, beyond which the Moon is pulling as the dominant gravitational force. It can be seen here that if the accident at 55:55 GET had not disabled the Service Propulsion System (SPS), a maximum burn from that engine would have had the crew back in 30 hours. *(NASA-MPAD)*

and time for Lovell and Swigert to get some rest, a six-hour period beginning at 70:00 GET according to the hastily rewritten flight plan, while Fred Haise stayed on watch. All were to be 'on duty' by 76:00 GET for flying around the Moon and getting set up for the PC+2 burn.

No free lunch

Unbeknown to all three astronauts aboard Apollo 13, in Mission Control one of the most crucial decisions of the rescue plan had already been decided. While everyone agreed the logic of conducting a PC+2 burn to cut the trans-Earth time, there was still a decision pending at the highest level about what to do regarding a superfast follow-up burn with the Service Module jettisoned. This impasse stretched the immoveable line separating senior management at the Manned Space Center from decisions made in the MOCR, where operational judgements were exclusively the prerogative of the duty flight director.

But this was no ordinary mission and a consensus of viewpoints resulted in uncertainties regarding the logic of optional paths to getting the crew back on Earth. A unique meeting was called by Chris Kraft, who summoned key people from the MOCR, involving Windler and Griffin, Gene Kranz from Room 210, top brass such as Administrator Paine, Center Director Gilruth, George Low the new deputy administrator, as well as Jim McDivitt and several assistants. Low had been with NASA since its formation 12 years earlier and he had masterminded recovery from the Apollo fire.

Unlike the deliberations under way in Room 210, this meeting was in full view of Jerry Griffin's flight controllers, conducted in the VIP viewing area behind and above the MOCR. Separated and soundproofed by a large glass screen, it was a comfortable area where a series of stepped seats gave it the feeling of a small cinema, a place where wives and families of the astronauts could watch the men watching the spacecraft. Now it was where the top echelon of NASA management met to hear options and recommendations.

First, Kraft wanted to get Flight Directors Lunney, Griffin and Windler together in one of the small support rooms just off the corridor from the MOCR, so Griffin handed over to his assistant and joined Kraft. Having decided a position, Kraft led the three men across to the VIP area to join the top brass. Supported by Charles 'Chuck' Deiterich (FIDO) and David 'Dave' Reed (RETRO) from his Gold Team, Jerry Griffin argued the case for keeping the Service Module attached and not exhausting virtually all the descent propulsion system propellant on one banzai burn.

Griffin's number-crunchers agreed with Kranz, fresh up from Room 210, that a balanced, calm and disciplined approach to the risk was preferable to an action that could leave the crew without reserves for course corrections vital to compensate for errors in the burn or the spacecraft attitude. However small those residuals, after travelling back from the vicinity of the Moon they could magnify to a catastrophic miss without the means to correct the trajectory. George Low could sympathise with the reasoning. He had introduced, as an early convert, the concept of Failure Mode & Effects Analysis (FMEA) that had got Apollo back on track after the fire of January 1967.

FMEA is a manifestation of inductive reasoning, a way to project forward for effective assurance of counterbalancing a sequence of failed elements, either systems or components. Applied to Apollo 13 it was a sound method for evaluating the probabilities of success or failure with any one mission mode choice, leaving deductive reasoning, a retrospective analysis of the failure, until after the crew got home. This level of reasoning left no argument over which path to choose: withdraw from the brink on both consumables and propellant conservation to lower the balance of risk on both without loading excess risk on one at the expense of the other.

But that was not the way the meeting went at first. In front of the NASA hierarchy, Griffin led the discussion, playing 'devil's advocate', seeming at first to support the superfast, all-or-nothing burn to get the men back fast. For his part, Deke Slayton supported that and pushed hard for a maximum burn dramatically shortening the time home. But Gene Kranz brought a cool logic not born from sophisticated systems management methodology but from running almost every mission since the first. Supporting him was Bill Peters, the TELMU from the Gold Team, a seasoned veteran of 16 manned space missions, who had worked for several hours to thrash out consumables.

The decision was a natural progression of logical thinking and it got the approval of senior NASA management. It had to be that way because, with caution delaying the return of Lovell, Swigert and Haise, if anything else went wrong someone very high up was going to have to tell the world why they had opted for the second slowest way back to Earth – an option, nevertheless, that would preserve equitable levels of consumables, including propellant. But in choosing to trade one advantage by paying the price of another it proved that an old maxim held good for space flight: there was no free lunch. It was a zero-sum game.

At 70:24 GET (11:35am in Houston), the public affairs officer announced to the world that the decision had been made to do the PC+2 manoeuvre with a modest burn of the descent engine, which would have the crew back at 142 hours. But for that to take place there had to be some certainty that the attitude of *Aquarius* was sufficiently well known to ensure that the thrust line of the descent engine, itself aligned through the centre of mass, would be pointing in the right direction. If not, it would undo all the meticulous planning and push the trajectory away from one that would bring the spacecraft back to Earth.

The MCC-2 burn to put Apollo 13 back on a free-return course had been made with the REFSMMAT (Reference For Stable Member Matrix, a mathematical method of determining a spacecraft's celestial orientation) from *Odyssey*'s computer transferred to *Aquarius*. Made less than four hours later, that burn had been brief, and the dispersions allowed were relatively large. Now, 18 hours further on, the PC+2 burn would be of much greater magnitude and the allowable dispersions were tighter. The standard procedure would be to do a navigation update through a P52 using separate star alignments to get a set of sightings, necessitating expenditure of RCS propellant to change the attitude of the spacecraft and electrical power too.

BELOW LEFT Jim Lovell 'poses' for a picture in *Aquarius*. *(NASA)*

BELOW Lovell's hand controller attached to the retractable armrest. *(NASA)*

Because the crew could not see stars with or without the AOT it was decided to recommend a Sun check before going behind the Moon to determine whether or not a P52 would be needed after all. The tolerable pointing error was within +/-1°, and if it was found possible to confirm that *Aquarius* was showing an accuracy within that band it would not be necessary to resort to a full navigation update. At 72:28 GET, with Lovell and Swigert resting, Charlie Duke took over the microphone from Capcom Vance Brand to explain to Fred Haise a scheme, evaluated by Young and Duke in the simulator, which would allow the AOT to be aligned on the Sun as if the operator was taking marks for a star sighting.

The Sun subtends an angle of 0.5°, which indicated that if the centre of the AOT was off by more than two Sun diameters there would be an equivalent platform misalignment of 1°. This Sun check would show if the Lunar Guidance Computer could place the AOT reticle within that deadband. Kerwin explained the procedure to Haise, but Lovell was already up, unable to get any real rest in *Odyssey*, which was getting colder by the hour. Also read to the crew was the procedure to get the rendezvous radar antenna out of the way for an AOT

alignment. In the stowed position it obstructed the optical path of the AOT, and was moved at 73:32 GET.

The AOT Sun check was completed at 73:47 GET with excellent results, the alignment being within 0.66 solar diameters, reassuringly displaying that the platform was off by a mere 0.33°.

Around this time Kranz relieved Griffin and brought his people back in to the MOCR to handle the flight around the Moon, the PC+2 burn and the subsequent deep power-down that would save consumables and ensure a timely return. What they had been doing in Room 210 was more than budgeting consumables and planning the magnitude of the PC+2 burn. All space operations require ground rules, essentially general statements which define what should, or should not, be done under expected and unexpected circumstances. Now, the totally unique situation into which the mission was propelled required new ground rules for the completely new sequences being planned.

Charlie Duke read up the complex sequence of pre-burn configurations for the crew to copy down on the checklists, flight plan pages and the activation manuals. At 75:28 GET, he read up the preliminary manoeuvre pad for the PC+2 burn, the Kranz team now fully installed in the MOCR, Griffin's men having joined Windler's team at the back of the room. One thing Kranz scrubbed was a back-up star check as Apollo 13 slipped into sunset prior to rounding the far side of the Moon.

With the glare of the Sun gone, lessening reflections off the myriad tiny particles escorting the damaged spacecraft, in theory it would be the best time to try to get star alignments. But it was an added pressure on time and would compromise the all-important PC+2 burn if complexities developed during the procedure.

As the Sun converged on the limb of the Moon, the crescent got thinner and the shadows allowed the crew to begin an experience that for two of them was to be a singularly unique event – a flight around the Moon totally out of sight of the Earth. Jim Lovell had been this way before but it was the only trip that Haise and Swigert would ever make.

Just before sunset came at 76:32:45 GET, Jim Lovell found a sight to momentarily take

his attention from the gloom of the situation when, seeing the sky full of lights, he gave the ecstatic cry: 'Man, look at those stars!' And then, through the triangular shaped windows of *Aquarius*, at his left forward position, Lovell could distinguish several stars that he recognised and would have given much to have been able to see earlier. There before them were Antares, Nunki, Scorpio and others with which they were familiar from the simulators and navigation updates.

Inside *Aquarius* the sunlight that had been with them since leaving Earth was gone and the crew reached for penlight torches to read written notes with scrawled settings and switch positions meticulously copied down. It was an ethereal moment for Haise and Swigert, a sad reflection for Lovell on what might have been on this, his second trip to the Moon – the first man

LEFT George Low had helped set up NASA out of the old National Advisory Committee for Aeronautics in 1958. Now, having helped forge NASA's great successes, he was deputy administrator under Thomas O. Paine. His role in Apollo 13 was supportive, and key to holding the upper echelons of management together at this time of crisis. *(NASA)*

LEFT Thomas Paine succeeded NASA Administrator James E. Webb when the latter retired as boss of the space agency in October 1968. He assumed office just in time to make the bold decision to send Apollo 8, only the second manned Apollo flight, on the first manned Saturn V all the way into Moon orbit and back. Jim Lovell was on that flight, and now he was the first astronaut to go back. It was his destiny to command the only Apollo mission to go to the Moon and not enter orbit. *(NASA)*

to go back a second time. As Apollo 13 homed in on a fast pass across the Moon a mere 156 miles (252km) above the barren surface, the crew got the cameras out in preparation for the spectacle which would unfold below.

Apollo 13 disappeared around the left side of the Moon as viewed from Earth at 77:08:35 GET. For several hours the spacecraft had been pulled at quickening speed toward the Moon, moving now at more than 4,800mph (7,723kph). In Houston it was 6:21pm and for the first time in three days nobody could speak to the crew. For a little over eight minutes after loss of signal (LOS) the crew was in total darkness, lit only by the wafting fingers of hand-held penlights.

As rapidly as it went the Sun reappeared, a flash of light across the lunar horizon extending probing fingers of brightness through the rutted and pockmarked undulations. Fred Haise saw it first and grabbed his camera. *Aquarius* had two 70mm Hasselblad data cameras with 60mm Zeiss Metric lenses on board and *Odyssey* had two similar cameras but with 80mm lenses, each with slightly different apparatus. Only two cameras were used for taking pictures, although in all, throughout the mission, some 584 photos were taken on five rolls of film with most of them showing the Moon.

For some time, alone in their crippled spacecraft, the Apollo 13 crew gazed at the surface and remarked on its utter desolation. Lovell slipped back and let Swigert have priority at the left triangular window. Below lay a place they would never get to visit again, full

LEFT Elements of the Environmental Control System (ECS), now so important for maintaining the life of the crew, with oxygen and water tanks indicated on Ascent and Descent Stages. Water was the most valuable and worrisome consumable, usage increasing with demands from the coolant system as systems were switched on, decreasing again after power-down. With critical events coming up, both battery and water consumption was high. *(Grumman)*

of craters, rolling hillocks and sinuous rilles. As the strength of the sunlight grew these alien features were washed in a brilliance that cast reflected light from the grey surface, texture and hue changing colour as the increasing angle of the light turned it to an ashen white.

Coming home

At 77:33:10 GET, just 24min 35sec after losing contact with Earth, Apollo 13 rounded the eastern side of the Moon – AOS, acquisition of signal – with a cheery 'Good morning Houston, how do you read?' from Jim Lovell. In fact, in Houston it was 6:44pm. As if drawn from the crisis of the mission, all three crew members continued discussing what they were watching far below, describing features they had studied for months, navigation points across the lunar surface, more familiar to these men than countries are to most Earthlings.

But it was quickly back to business, with a disciplined jolt for one crew member at least. Ever the Commander, Lovell turned to switch settings and preparations for the power-up prior to the upcoming engine burn that would hurry them home. For several minutes Haise and Swigert were distracted, captive to the view from the two windows, the conical slope of the Command Module and the RCS clusters on their support brackets feeding in to a stunning view of the Moon, still very close, rolling slowly by. A sharp reminder from Lovell and his crew were back.

Within five minutes of rounding the Moon and coming back into contact, Apollo 13 was already 650 miles (1,046km) above the surface and slowly pulling away at a speed of 4,815mph (7,747kph), slowly decreasing with time as the Moon tried to pull it back before once again the Earth's gravity field would become dominant and it would begin to speed up.

Before reading up the numbers and values for the P40 manoeuvre, there was a final confirmation from Brand at the Capcom console that there would be no additional refinements of the alignment: 'Before I start on the pads, another comment. The general indications that we gave you before about the Sun being in the Commander's window and about stars in the AOT, such as might be used as general

indication for your attitude for the burn, that's all out the window. We are just going on the Sun check that we made earlier...'

And so the pre-burn preparations were denied one more tiresome and potentially frustrating procedure. Vance Brand read up the P40 for Lovell and Haise to punch into the PGNCS via the keyboard, with the precise time, magnitude of the burn, delta-velocities in all three axes and the calculated weights of the two spacecraft: 62,480lb (28,341kg) for *Odyssey* and 33,452lb (15,174kg) for *Aquarius*. Your writer was tasked with calculating the weight at the time of ignition, and resolved a value for *Aquarius* of 33,392lb (15,147kg). In the bizarre world of programmed burn calculations, the 'current' mass was the benchmark.

The P40 also defined the manually controlled throttle settings: 5sec at minimum thrust, 21sec at 40% and the remaining time, about 238sec,

BELOW The mid-section contained most of the environmental control equipment aboard *Aquarius*, with water and oxygen tanks outside the pressurised area. This schematic shows the relative locations of ECS equipment and the control modules for respective subsystems. *(Grumman)*

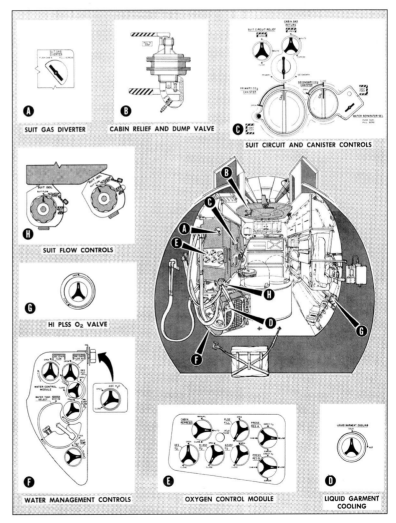

A SUIT GAS DIVERTER

B CABIN RELIEF AND DUMP VALVE

C SUIT CIRCUIT AND CANISTER CONTROLS

H SUIT FLOW CONTROLS

G HI PLSS O₂ VALVE

F WATER MANAGEMENT CONTROLS

E OXYGEN CONTROL MODULE

D LIQUID GARMENT COOLING

THE BIG HURRY-UP

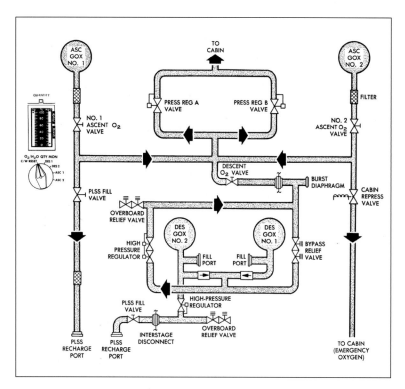

ABOVE The oxygen-supply and cabin-pressure control system integrated both Descent and Ascent Stage supplies. Oxygen was not going to be a problem, lasting far beyond the most pessimistic predictions for the estimated time of re-entry. *(Grumman)*

BELOW This page from the Apollo 13 Flight Plan shows the plan to place the spacecraft in lunar orbit, which for this mission would not happen. It does, however, show the orientation of the Moon as Apollo 13 approached it for a fast fly-by, disappearing around the western limb and appearing around the eastern side 24 minutes later. *(NASA-MPAD)*

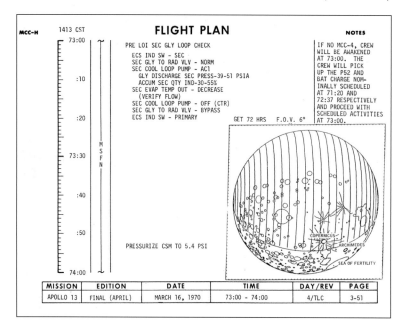

at full throttle. This way, by starting at a low throttle setting, if anything was wrong with the engine it would hopefully manifest itself at a relatively low velocity. If the engine shut down prematurely the dispersion from the nominal trajectory could be calculated to within the ΔV of the RCS thrusters. But there were some concerns about the descent engine.

The gap of 16 hours between firings resurrected old fears that the engine might not be stable the second time. And then there was the worry about the SHe tank and the increasing pressure. If the burst disc popped there would be no means of pressurising the propellant tanks and no descent engine. And if that happened the mission would have to survive another ten hours.

There was, of course, another main engine in *Aquarius*. But there was no purpose in jettisoning the Descent Stage to use the Ascent Propulsion System (APS), because that stage held the oxygen, water tanks and batteries vital for keeping the crew alive. Just as the Service Module was designed to serve as the engine room and consumables store for the Command Module, so the Descent Stage provided that function for the pressurised Ascent Stage. And just so that some number could be quoted to define that fact, your writer was also asked to calculate that.

At the very time Fred Haise was reading back to Vance Brand the P40 pad for the PC+2 burn (77:56 GET) the third stage of the Saturn V slammed into the Moon as planned. It produced shock waves through the surface picked up by the seismometer left by Apollo 12 astronauts Conrad and Bean five months earlier, telling scientists on Earth a little bit more about the Moon's interior. Six minutes later Brand informed the crew: 'By the way, *Aquarius*, we see the results now from 12's seismometer. Looks like your booster just hit the Moon, and it's rocking it a little bit. Over.'

With not a little tinge of frustration at a mission turned sour, Jack Swigert dismissed the comment with a laconic swipe: 'Well, at least something worked on this flight.' And referring to the plan that would have sent the Ascent Stage too spiralling down to a lunar impact after returning its crew to *Odyssey* after more than 33 hours on the surface: 'I'm sure glad we didn't have a LM impact too!' And then it was back to business with an advisory from the ground that power-up for the burn would begin at 78:12 GET.

On through the next hour or so the crew and the MOCR worked up the appropriate switch positions and procedures for the all-important burn. With ten minutes remaining, Gene Kranz went around the room calling up his flight controllers for a 'go/no go' decision. With the clock ticking down the MOCR began to fill up, the Apollo 14 crew of Alan Shepard, Stuart Roosa and Edgar Mitchell down among the trenches, the VIP room populated now with NASA top management and some senior astronauts including Frank Borman, commander of Apollo 8.

The hurry-up burn two hours after pericynthion saw ignition right on time at 79:27:38.3 GET, preceded by a 10sec burn of two RCS jets to settle the propellants in an ullage manoeuvre. In Houston it was 8:40pm. The descent engine fired for 4min 23.82sec for a ΔV of 860.5ft/sec (262.1m/sec), or 586.69mph (943.98kph). The burn had consumed 7,882lb (3,575kg) of propellant, leaving *Aquarius* with a post-burn mass of 25,463lb (11,550kg). At shutdown Apollo 13 was travelling at 5,020.2ft/sec (1,530.2m/sec) or 3,422.8mph (5,507.3kph). The Moon was 6,511.2 miles (10,476km) distant and getting smaller.

The residuals from the burn (a quotient of the errors) were close to zero, a remarkable achievement given the circumstances. Subsequent tracking showed that this would bring the spacecraft no closer to the Earth than 100.6 miles (161.8km), too high to be captured by the atmosphere, so a further course correction would be necessary, albeit minor. As it was, Apollo 13 was still slowing down, and would continue to be slowed by the Moon's gravity for almost 11 hours until it passed the point where the gravity fields of Moon and Earth balance (equigravisphere), beyond which they would start to speed up.

Immediately after the burn, key players converged on the flight director's console to discuss the next events. Deke Slayton argued for getting the crew to sleep – immediately; Chris Kraft wanted to get the power-down under way – fast; and spacecraft designer

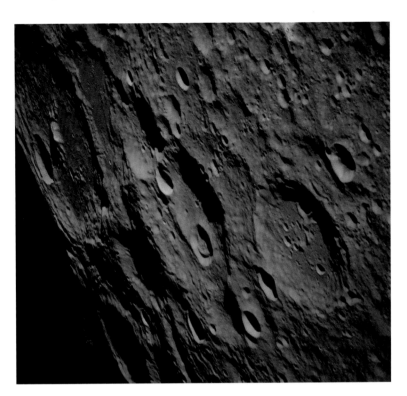

ABOVE As *Aquarius,* with *Odyssey* in tow, slipped around the west limb of the Moon, the surface was in shadow and would only emerge into Sunlight as the spacecraft passed around the far side. *(NASA)*

BELOW To Jim Lovell these were familiar features. He had been this way before, accompanied that time by Frank Borman and Bill Anders only 16 months earlier, during the epic flight of Apollo 8, the first mission to leave Earth's gravity and travel to another world in space. *(NASA)*

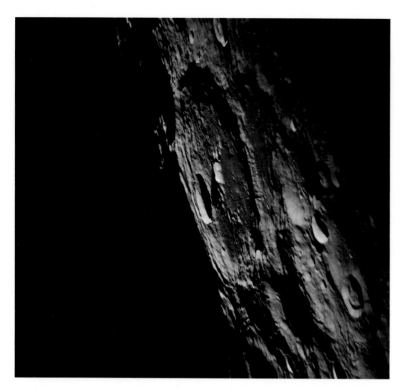

ABOVE The dark shadows across the crater Chaplygin, 6.2°S by 150.3°E. With a diameter of 85 miles (137km), it lies on the far side of the Moon as viewed from Earth. *(NASA)*

BELOW The two major craters in this photograph are Chaplygin (just left of centre), and Schliemann (below centre), viewed through the windows in *Aquarius*. *(NASA)*

Max Faget, the brain behind NASA's manned vehicles to date, wanted to set up a PTC to alleviate any thermal pressure on the spacecraft – straight away. All these requirements were mutually exclusive, so Kranz made the decision to set up the PTC and power down *Aquarius* and then allot some much-needed sleep for the crew, in that order.

Turning off the lights

At 80:00 GET the situation with consumables looked bleak. The PC+2 burn ensured that the Command Module would be on the water at 142:40 GET. But as it stood, 30 minutes after the hurry-up burn, the water supply was running at 4.8lb/hr, with depletion projected at 126:06 GET. The oxygen consumption rate was down and would stretch to 253:00 GET, but at current levels of consumption at 37Ah the battery power would be gone by 120:48 GET. Unresolved but being worked on, the lithium hydroxide canisters from *Odyssey* would somehow have to be strapped on to the hoses for *Aquarius* to reduce carbon dioxide levels.

At no time since the crew powered up *Aquarius* at 58:40 GET had the consumption of electrical power dropped below 24.9Ah. The use of LM systems, especially the PGNCS and other guidance and navigation equipment, had taken its toll, reaching 37.6Ah during PC+2. Now, in the subsequent power-down, use would have to fall to 10–12Ah. Because of the link between electrical power and water consumption for the cooling system, reduced power levels would lower temperatures and cut water consumption from 4.8lb/hr (2.18kg/hr) to 2.5lb/hr (1.13kg/hr), taking it out to 150:00 GET.

As things stood at this point, *Aquarius* would have to be powered up about 35–40Ah around 134:00 GET to top up the *Odyssey*'s Command Module batteries for re-entry after *Aquarius* was cut loose, and for supplying power to the CM, getting it ready to support the crew when it was finally on its own shortly before entering the atmosphere. In summary, the moderate power levels since first getting into *Aquarius*, and the high power levels of the two engine burns, would be followed now by a very deep cut in electrical use for the next 52 hours before finally turning on the systems again prior to re-entry.

At NASA facilities and among the host of contractors and subcontractors now working across the country to solve myriad minor issues regarding the consumables, a team at the Windsor Locks facility of Hamilton Standard were fathoming out whether they could use the 14.5 hours of water remaining in the Command Module to supplement the supply in *Aquarius*. The problem was that the water in *Odyssey* was chlorinated and that in *Aquarius'* environmental control system was not. Could the chlorinated water be used in the LM's sublimator? Tests showed that it could.

Then Hamilton Standard's backpack manager Cal Beggs worked up a method whereby the crew could transfer this chlorinated water from *Odyssey* into the supply in *Aquarius*. The Personal Life Support System or PLSS (pronounced 'pliss') backpacks were pre-charged with six hours of water supply for the Moonwalks. A team including James Morancey worked up a procedure whereby the PLSS water could be fed into the main supply in *Aquarius*, reversing the normal procedure for refilling each PLSS for a second Moonwalk. Then they found a way to use the empty tanks to ferry water from *Odyssey*, but with water consumption slashed these back-up techniques were not needed, although the engineers even considered using urine to keep the coolant system going!

Aboard *Aquarius* and in conjunction with ground operations, power-down began within five minutes of the end of the PC+2 burn, and at 79:52 GET the ground read up a fairly complex procedure for setting up a passive thermal control mode. But problems with stabilising the barbecue roll persisted for more than an hour until finally resolved. During this period the crew reported seeing sizeable chunks of debris floating alongside the spacecraft, possibly material loosened from the crippled Service Module by the vibrations of the PC+2 burn.

As Mission Control eventually knew it would, the CO_2 level reached 7.5mmHg at 80:30 GET, triggering a master alarm, a level set pre-flight to advise the crew to change the LiOH canister. But the medical team advised the crew that it could safely run up to a concentration of 15mmHg without any noticeable effect and that when it

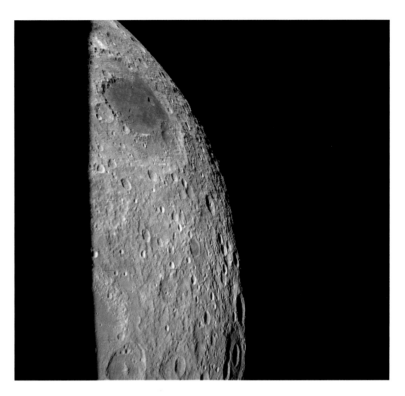

ABOVE Mare Moscoviense (Sea of Moscow), at 27.3°N by 147.9°E, is the remains of a basaltic flood plain 172 miles (277km) in diameter. *(NASA)*

BELOW Still around the far side of the Moon, but coming further towards the eastern limb, Apollo 13 passes close to the crater Tsiolkovsky, 112 miles (180km) in diameter at 20.4°S by 129.1°E. *(NASA)*

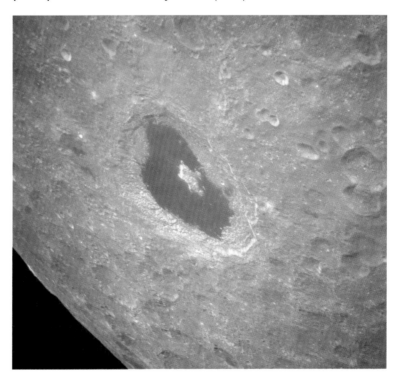

reached that level in about five hours they should change the primary canister for the secondary. After that the special procedure for using *Odyssey*'s canisters could be implemented. In tests a level of 22mmHg could be tolerated, but nobody wanted to go that high.

As related earlier, there were not enough round-shaped canisters to last, and the more prolific square-shaped cartridges in *Odyssey*

would have to be used in *Aquarius* – somehow! By late Tuesday, 14 April, as Lovell and Swigert were shortly to begin their sleep period, Beggs and around 100 people at Hamilton Standard had been working on a solution and kept up an intensive dialogue with the NASA people in Mission Control to get it validated.

The original idea to construct a makeshift adapter allowing the canisters from *Odyssey* to operate in *Aquarius* had been born in the mind of a NASA engineer who had lived and worked with the LM Environmental Control System (ECS) for several years. As chief of the Crew Systems Division, Ed Smylie, together with his assistant Jim Correale, had been responsible for managing the design of all equipment related to crew activity, and from within hours of the news about the accident he had concocted a way of keeping the atmosphere clean. Working with the environmental test chamber in Building 7, Smylie and Correale had duplicate LiOH canisters flown from Cape Canaveral to Houston, and by late afternoon on 14 April they had rigged up the test run in Building 7 and demonstrated that the system worked.

The crew would use tape and stiff card, together with plastic bags and ingenuity, to stack the square canisters from *Odyssey* and

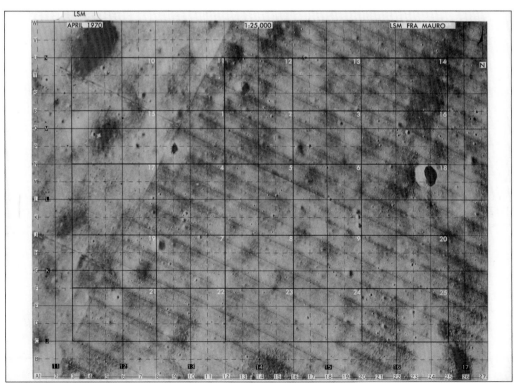

attach them to hoses leading from the ECS in *Aquarius* on connections normally used to hook up the astronauts' space suits. In this jury-rigged concept, contaminated oxygen would be drawn into the square cartridges and scrubbed of CO_2 before circulating back into the LM ECS and fed out via a hose attached to the outlet that would normally be used to clean oxygen from the suit.

In this rather complex system of boxes, hoses and pipes, the LiOH canister receptacle would be bypassed, the oxygen scrubbed of contaminating carbon dioxide outside, rather than inside, the ECS. A second inlet-outlet arrangement would be set up in *Odyssey* by stretching hoses through the tunnel from *Aquarius*. At least, that was the idea. Smylie brought his 'mailbox', as it became known, up to Mission Control to show the flight controllers what was needed. It would now have to be built by the crew from sundry items found around *Aquarius* and *Odyssey*, but not just yet.

At 80:37 GET (9:50pm in Houston), Vance Brand read up to the crew the basic idea and what was needed, in notably non-technical language: 'We have a lengthy procedure here, but in short you use plastic as a covering for the whole thing. You put some kind of stiffener against the LiOH entrance side. You need grey tape to stick the whole thing together and you need something like a sock to put in the bottom so that the outlet side is plugged up. As it turns out the flow is rather U-shaped through the cartridge...If you plug up the bottom it comes in one side of the top and goes out the other.'

Within 45 minutes Brand suggested the crew get something to eat before Lovell and Swigert began a rest period at around 82:00 GET, current power levels now coming down toward 12A, which would be achieved within three hours, lower than anything seen since the accident. A simple calculation ($P[W]=I[A]xV[v]$) shows that with a 28V DC current from the LM batteries, a 12Ah power usage produces around 336W. Such was the level of power running *Aquarius* – less than the power from four ceiling light bulbs in the average home; one third of the power needed to run a one-bar electric heater or a toaster!

At 81:46 GET, Brand read up a report that trajectory analysis after the PC+2 burn indicated

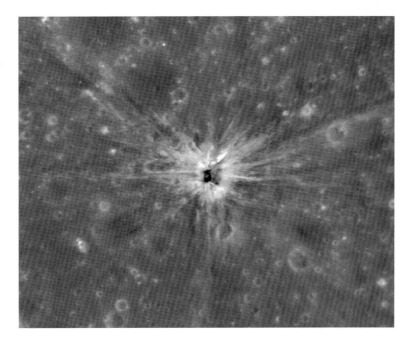

ABOVE The third stage of the Saturn rocket struck the lunar surface as planned, 23 minutes after Apollo 13 reappeared around the eastern limb, creating an impact that sent shock waves around the Moon. *(NASA)*

BELOW The Apollo Lunar Surface Experiments Package (ALSEP) contained a passive seismometer left on the surface by Apollo 12 the previous November, which recorded the seismic waves from the impact of the S-IVB stage, providing science data on subsurface structure of the Moon's outer layers. *(NASA)*

a need for a further midcourse burn of 4–6ft/sec (1.22–1.83m/sec) using the descent engine, and gave Lovell a detailed rundown on the consumables status – what remained, how deep the power-down had to be and how long each of the prime assets would last. After that review it was time for Gene Kranz to step down, relinquish the console to Milt Windler and his Maroon Team and get back to Room 210. That got under way at 82:00 GET – 11:13pm in Houston.

Before the shift change, Brand passed up a further message: 'Deke [Slayton] says get a night's sleep. He says you've been working hard and you ought to relax a little bit and be ready for tomorrow.' It was coming up on midnight, a new day about to dawn – Wednesday 15 April. By midnight Swigert was finishing up a meal and Haise was completing final power-down steps, while Lovell was already up in *Odyssey* preparing his bunk for the night.

On the stroke of midnight Houston time, Haise got on to Jack Lousma at the Capcom console: 'OK, when I was upstairs [in *Odyssey*] just a minute ago I noticed what appeared to be some new venting from down the Service Module way. I noticed that out window 1, and also saw one chink of metal – loose metal about four inches [10cm] square that was

STAR / PLANET LIST

OCTAL STAR CODE	NAME	VIS. MAG.	RIGHT ASCENSION (HR. MIN.)		DECLINATION (DEG. MIN.)	
1	ALPHA ANDROMEDAE (ALPHERATZ)	2.1	0	06	+28	53
2	BETA CETI (DIPHDA)	2.2	0	42	-18	11
3	GAMMA CASSIOPEIAE (NAVI)	2.2	0	54	+60	27
4	ALPHA ERIDANI (ACHERNAR)	0.6	1	36	-57	25
5	ALPHA URSAE MINORIS (POLARIS)	2.1	1	58	+89	06
6	THETA ERIDANI (ACAMAR)	3.4	2	57	-40	26
7	ALPHA CETI (MENKAR)	2.8	3	00	+03	56
10	ALPHA PERSEI (MIRFAK)	1.9	3	22	+49	44
11	ALPHA TAURI (ALDEBARAN)	1.1	4	34	+16	26
12	BETA ORIONIS (RIGEL)	0.3	5	12	-08	15
13	ALPHA AURIGAE (CAPELLA)	0.2	5	13	+45	57
14	ALPHA CARINAE (CANOPUS)	-0.9	6	23	-52	40
15	ALPHA CANIS MAJORIS (SIRIUS)	-1.6	6	44	-16	40
16	ALPHA CANIS MINORIS (PROCYON)	0.5	7	37	+05	19
17	GAMMA VELORIUM (REGOR)	1.9	8	08	-47	14
20	IOTA URSAE MAJORIS (DNOCES)	3.1	8	50	+48	30
21	ALPHA HYDRAE (ALPHARD)	2.2	9	26	-08	30
22	ALPHA LEONIS (REGULUS)	1.3	10	06	+12	09
23	BETA LEONIS (DENEBOLA)	2.2	11	47	+14	46
24	GAMMA CORVI (GIENAH)	2.8	12	13	-17	20
25	ALPHA CRUCIS (ACRUX)	1.6	12	24	-62	49
26	ALPHA VIRGINIS (SPICA)	1.2	13	23	-10	58
27	ETA URSAE MAJORIS (ALKAID)	1.9	13	46	+49	30
30	THETA CENTAURI (MENKENT)	2.3	14	04	-36	11
31	ALPHA BOOTIS (ARCTURUS)	0.2	14	14	+19	22
32	ALPHA CORONAE BOREALIS (ALPHECCA)	2.3	15	33	+26	50
33	ALPHA SCORPII (ANTARES)	1.2	16	27	-26	21
34	ALPHA TRIANGULI AUSTR. (ATRIA)	1.9	16	43	-68	56
35	ALPHA OPHIUCHI (RASALHAGUE)	2.1	17	33	+12	35
36	ALPHA LYRAE (VEGA)	0.1	18	36	+38	45
37	SIGMA SAGITTARII (NUNKI)	2.1	18	53	-26	20
40	ALPHA AQUILAE (ALTAIR)	0.9	19	49	+08	46
41	BETA CAPRICORNI (DABIH)	3.2	20	19	-14	54
42	ALPHA PAVONIS (PEACOCK)	2.1	20	23	-56	51
43	ALPHA CYGNI (DENEB)	1.3	20	40	+45	09
44	EPSILON PEGASI (ENIF)	2.5	21	42	+09	42
45	ALPHA PICIS AUSTRINUS (FOMALHAUT)	1.3	22	56	-29	49
46	SUN					
47	EARTH					
50	MOON					
00	PLANET					

STAR	NO.	STAR	NO.	STAR	NO.
ACAMAR	6	CANOPUS	14	MIRFAK	10
ACHERNAR	4	CAPELLA	13	NAVI	3
ACRUX	25	DABIH	41	NUNKI	37
ALDEBARAN	11	DENEB	43	PEACOCK	42
ALKAID	27	DENEBOLA	23	PROCYON	16
ALPHARD	21	DIPHDA	2	POLARIS	5
ALPHECCA	32	DNOCES	20	RASALHAGUE	35
ALPHERATZ	1	ENIF	44	REGOR	17
ALTAIR	40	FOMALHAUT	45	REGULUS	22
ANTARES	31	GIENAH	24	RIGEL	12
ARCTURUS	31	MENKAR	7	SIRIUS	15
ATRIA	34	MENKENT	30	SPICA	26
				VEGA	36

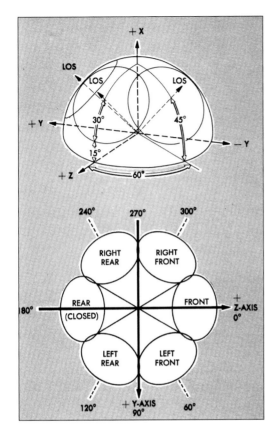

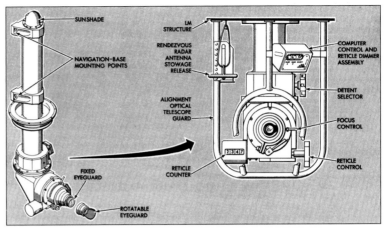

ABOVE The LM Alignment Optical Telescope (AOT) in *Aquarius* was used for star navigation sightings. Mounted to the navigation base, it was a periscope-type device with a 60° conical field of view, manually operated with a moveable shaft axis and a line of sight about 45° from the X-axis in the Y–Z plane. (*Grumman*)

ABOVE The LM Alignment Optical Telescope (AOT) in *Aquarius* was used for star navigation sightings. Mounted to the navigation base, it was a periscope-type device with a 60° conical field of view, manually operated with a moveable shaft axis and a line of sight about 45° from the X-axis in the Y–Z plane. (*Grumman*)

LEFT Detents and field of view from the AOT for celestial navigation. (*Grumman*)

BELOW A receding Moon as the crew begin their journey back to Earth, the distance diminishing with each passing hour. Now it was a matter of survival after powering down as many systems as possible to conserve battery power and water. (*NASA*)

tumbling around – silver in colour and it looked like it had come from somewhere down in the Service Module…I'm looking out the LM window now and I see a good part of the new star field it's created for us. There are about a thousand little sparklies out here.'

At 12:35am, Gene Kranz, astronaut Anthony England, INCO William Peters and CONTROL (control officer) Richard Thorsen hosted a press conference in the main auditorium in Building 1, chaired by public affairs officer John McLeaish. From the outset, the press had been invited into all activity, and two pool correspondents were allowed to sit in the MOCR and observe what was actually going on, reporting what they saw.

The first 24 hours after the accident had been focused on getting Apollo 13 back on a free-return trajectory, rounding the Moon and speeding up to cut the trans-Earth time by ten hours. The second 24 hours would be spent powered down to the absolute minimum, rigging up the makeshift 'mailbox' for keeping down CO_2 levels. The third 24 hours would be committed to refining the trajectory with a course correction and getting the procedures ready to power up the Command Module for re-entry.

Chapter Six

The long haul

Earth beckoned, but systems were creaking and could fail at any hour. Using as little electrical power as possible to conserve energy, the temperature was going down – close to freezing. And water was gathering on the walls, ice building up on the windows. It was now a matter of survival on the long haul back home. But would *Odyssey* be in a fit state to survive re-entry?

OPPOSITE The long ride home was only just beginning, with unresolved problems ahead as the crew settled in to a long, cold, damp ride back to Earth. *(NASA)*

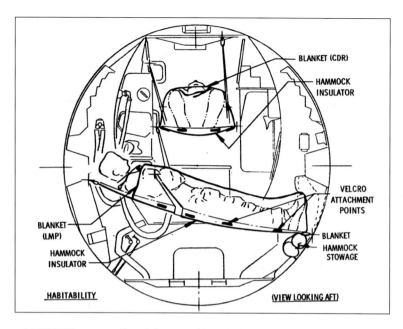

ABOVE The conventional sleep positions in *Aquarius* were akin to camping out, one man slung across the two forward station positions, the other in a hammock across the top of the mid-section. These positions were devised for sleeping in $\frac{1}{6}$th gravity while on the surface of the Moon. Now, the LM was a weightless campsite, operating in a manner never envisaged when designed. *(Grumman)*

BELOW In the dimly lit interior, some systems are up and running in this shot which shows an area between the two crew stations. To the left is the left-hand triangular, forward-facing, window. *(NASA)*

In Mission Control it was Wednesday 15 April, and in space Lovell and Swigert were in their respective couches in *Odyssey*, trying to keep warm, trying to get some sleep and trying to ignore Fred Haise down in *Aquarius* occasionally talking to the lone voice from Earth. Lousma kept up a conversation, keeping Haise well informed about the many astronauts working the simulators, including Ken Mattingly, who was conducting simulated power-up procedures for the final hours before entry.

Then it was time to discuss a wide range of camera settings for Fred to try and get some pictures of the 'sparklies', with the comment from *Aquarius* that the Moon was still so 'big and bright...I can just barely, on the left corner of the Moon now, make out the foothills of Fra Mauro formation. We never did get to see it when we were in close there...From the sounds of all the work that is going on and is still going on, this flight is probably a lot bigger test for the system on the ground than up here.'

Considering the pressure of the last 28 hours, the morale of the crew was high, but as the conversation between Fred Haise and Jack Lousma continued it was clear that these little gossip sessions had much value. Haise enthusiastically told Lousma about the 'beef and gravy and other assorted goodies' that he was about to 'tear into' and all the while the flight surgeon on duty listened intently for signs of stress or pressure.

Lousma asked about the CO_2 level and was told it was around 13mmHg, which he already knew, and chatted with Fred about the way the PTC seemed to be going off alignment. It meant nothing serious, but it was something on which to concentrate the mind.

It was quiet in Mission Control, the elapsed time clock ticking silently second by second. Outside it was a starry night and some off-duty controllers gathered to gaze toward the sky, unable to sleep, chatting for the first time in two days, puffing on cigarettes. In the MOCR, as the clock hit 1:47am and after four minutes of quiet on the communications line, Jack Lousma asked a very normal but a very human question: 'How are you feeling, Fred?'

'Oh, as soon as I chug down this grape drink and grapefruit – orange drink, I think I'll be in pretty good shape,' he replied.

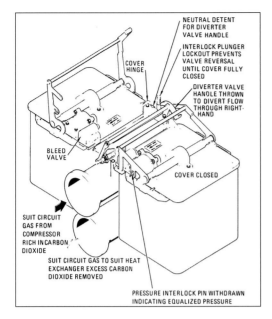

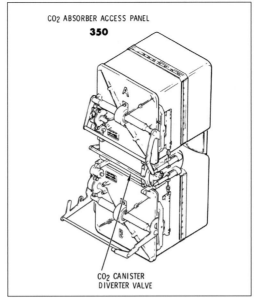

CO₂ ABSORBER ACCESS PANEL
350

CO₂ CANISTER
DIVERTER VALVE

'How much sleep did you get between the burn and the time you got up for this exercise?' The question was prompted by the flight surgeon, who had wandered over to have a word in Lousma's ear. CO_2 levels were higher now than they had ever been in a manned spacecraft. Concentration could be fading.

'Oh, I'd guestimate about four hours Jack. Wait a minute, which burn are you talking about?' said Haise.

'It was the burn we just made. Did you get any sleep between it, and the time you got up for this watch?' Lousma asked for clarification.

'Oh no, I didn't – I haven't been to bed since all the action the night before and I went to go to bed for about four hours then...'

'Well, we're just trying to figure out who's likely to be the most tired up there. You or Jim,' explained Lousma.

'I think we'll get caught up pretty good in the next couple of days,' said Haise confidently.

And so it went on, as Jack Lousma kept Fred talking until the time came to change from the primary LiOH canister to the secondary at 85:23 GET, with the CO_2 level reading 14.9mmHg. It began to fall rapidly almost immediately and there were still several hours to go before the 'mailbox' would be needed.

Jim Lovell finally stirred after little more than three hours' rest and floated back down into *Aquarius*, with Swigert still sleeping on his couch in *Odyssey*. It was 3:04am in Houston, 85:51 GET. Joe Kerwin was about to relieve Lousma

at the Capcom console while aboard Apollo 13 Fred Haise was getting his head down in *Odyssey*. On the ground and in space there was continuing attention to the PTC mode, which appeared to be coning rather than performing a stable 'barbecue roll'. The LM RCS system had never been designed for this purpose, and with two-thirds of the total docked mass hanging out on the end of the vehicle doing the attitude control, excursions were bound to build up.

TABLE 1-4

LIOH CANISTER CHANGE SCHEDULE

CHG. NO.	APPROX. GET HRS	APPROX. ΔT	INSTALL CAN	POSITION	REMOVE & STOW CAN NO.	STOWAGE LOCATION
1	9:00	14	3	A	1	B5
2	23:00	12	4	B	2	B5
3	35:00	13	5	A	3	B5
4	47:45	12	6	B	4	B5
5	60:00	11	7	A	5	B6
6	71:00	14	8	B	6	B6
7	85:15	13	9	A	7	B6
8	98:30	14	10	B	8	B6
9	112:40	29	11	A	9	A3
10	141:20	13	12	B	10	A3
11	154:30	12	13	A	11	A3
12	166:35	15	14	B	12	A3
13	181:15	11	15	A	13	A4
14	192:20	15	16	B	14	A4
15	207:30	9	17	A	15	A4
16	216:30	15	18	B	16	A4
17	231:50		19	A	17	A6

To help plot how the coning was progressing, Mission Control asked Haise to monitor the position of the Moon and the Earth using the Landing Point Designator (LPD) as each came into view across the windows. The LPD was a calibrated grid inscribed in the two window panes, designed to tell the pilot where the computer was taking the LM as it descended towards the surface. With only one position through which to view the two sets of inscribed numbers overlaying each other, the astronaut would line them up and look through the numbers to see the location on the surface. From that he would decide whether to take over manual control and redesignate the landing site.

As Fred read the position of the Moon or the Earth as each came across the LPD marks, he was able to provide Kerwin with a readout that gave Mission Control a measure of how the coning was progressing. With the docked vehicles making one complete revolution every 11 minutes the coning was increasing, and there was some discussion about that and the apparent venting of materials coming out of the Service Module, observed through the windows.

For the next several hours a running dialogue was maintained, reading up routine data pads for a course correction burn in the unexpected event of a communication failure. There was more discussion about the spacecraft alignment to achieve that, and still further concerns from Jim Lovell about the apparent venting from the Service Module. The venting did not appear to be affecting the trajectory, but a course

correction would be necessary to tweak the flight path back on to the correct angle of entry into the Earth's atmosphere.

Around 90:00 GET (7:13am Wednesday 15 April) Jack Swigert joined Jim Lovell in *Aquarius*, leaving Fred Haise asleep up in *Odyssey*. The Black Team were now coming on shift under Flight Director Glynn Lunney with Capcom Joe Kerwin. Milton Windler's Maroon Team were off busy preparing checklists and switch positions for the coming power-up of *Odyssey* and the course correction, and Lunney and some controllers were called away for that conference. From the MOCR there followed a lengthy read-up of the items necessary to build the jury-rigged 'mailbox' for getting CO_2 scrubbed using canisters from the Command Module and instructions on how to build the device.

The crew would use items previously scavenged from storage lockers in *Odyssey*: duct tape, cut into strips, an EVA Q-card which would have been used as a prompt for functions on the lunar surface, plastic bags for liquid cooled garments which would have been used for suit cooling on the Moon, scissors, a towel and more tape. For more than an hour the build instructions were passed to Apollo 13, while all three crew members built the first box for hooking up into the suit hoses masquerading as flow channels for the LiOH canister from *Odyssey*, a jury rig known to Texans as 'shade tree engineering'.

While all this was going on, Apollo 13 passed another milestone – the equigravisphere. At precisely 90:24:40 GET (7:37am in Houston), Apollo 13 was 38,921 miles (62,624km) from the Moon and 216,427 miles (348,232km) from the Earth, and would now begin to speed up. But aboard *Aquarius* there was work to do. After building the first 'mailbox', Lovell and Swigert were asked to build another, but to hold off on switching them on so as to get the most out of the secondary LiOH canister already in operation.

The working arrangement between the different flight control teams was now getting into a routine and a rhythmic flow: the three operational teams would confer about the tactical running of mission sequences based on hard strategies developed by Kranz's people in Room 210. Lunney questioned the use of PLSS

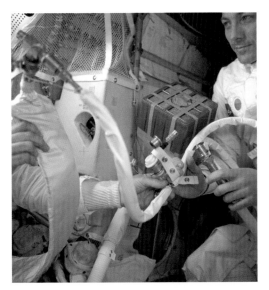

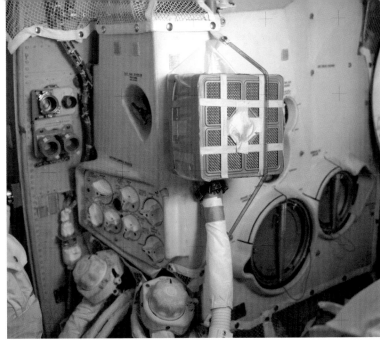

ABOVE LEFT The jury-rigged adapter for utilising *Odyssey*'s copious supply of lithium-hydroxide canisters in *Aquarius* comes together, as Fred Haise takes a picture of Lovell and Swigert doing a spot of do-it-yourself work! *(NASA)*

ABOVE One canister was attached to the side of the environmental control station in *Aquarius,* where it would scrub the oxygen in the LM. *(NASA)*

LEFT Using spare card, duct tape and some ingenuity, a workaround procedure devised on the ground was transformed into a smart unit for preventing asphyxiation, which would have occurred long before the crew got home. *(NASA)*

BELOW LEFT Neat and tidy, the first canister adapter is strapped to the side of *Aquarius.* Note the upper hatch to *Aquarius* is permanently open to permit a flow of oxygen up in to *Odyssey*. *(NASA)*

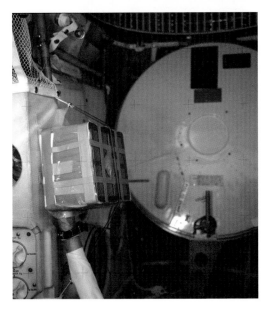

LEFT The second canister was placed up in the tunnel between *Aquarius* and *Odyssey.* Exhaled carbon dioxide could accumulate in the recesses of the Command Module unless it too was kept clean of contaminating gas. *(NASA)*

water and whether the ascent tanks should be back-fuelled with water from these Moonwalk packs. It was decided not to invoke the added complexity of draining the backpacks because there might not be enough delta pressure to get a positive flow. He also questioned whether the Descent Stage tank water would freeze by 140:00 GET.

Then there was discussion about the course correction burn (MCC-5), with a judgement that it should be done sooner rather than later, one rationale for which was the impending rupture of the burst disc in the supercritical helium tank used for pressurising the descent engine. If the disc burst before what was being referred to

as MCC-5, the burn could not take place. The SHe disc was calculated to relieve itself around 107:00 GET, so the burn was scheduled for two hours earlier.

Back in the MOCR the bootstrap 'mailbox' procedure had resulted in two good boxes for use when necessary, and at 91:53 GET Jack Swigert began copying down a torrent of new switch configurations. Fifty minutes later, with Fred Haise awake and talking to Houston about the condition of a device known as the ASA – the Abort Sensor Assembly – he floated down into *Odyssey* and set up the Command Module panels as directed, necessary for powering up *Odyssey* from the LM batteries later on.

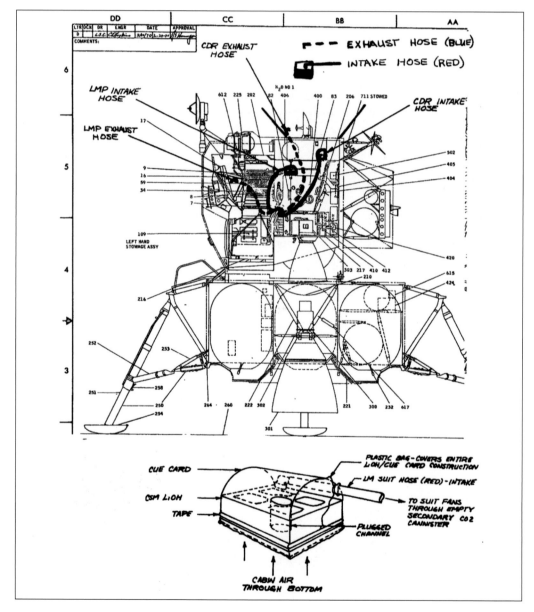

RIGHT A drawing from Grumman, worked on by systems specialists in Houston to reconfigure the oxygen flow and CO_2 cleansing. This alone was the key to allowing the LM to serve as a lifeboat, and was the most important contribution made by the crew to their survival. (Grumman)

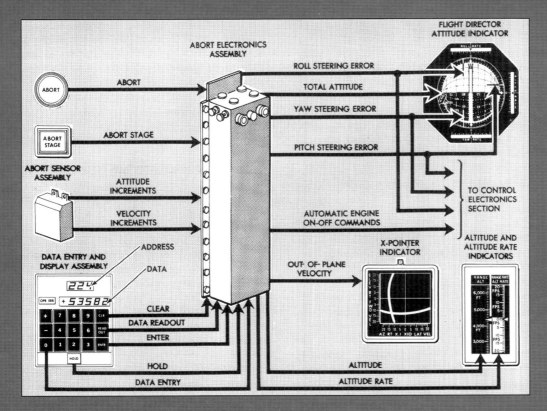

The following labels appear on the diagram:

FLIGHT DIRECTOR ATTITUDE INDICATOR

ABORT ELECTRONICS ASSEMBLY

ABORT

ROLL STEERING ERROR

TOTAL ATTITUDE

ABORT STAGE

YAW STEERING ERROR

PITCH STEERING ERROR

ABORT SENSOR ASSEMBLY

ATTITUDE INCREMENTS

VELOCITY INCREMENTS

AUTOMATIC ENGINE ON-OFF COMMANDS

TO CONTROL ELECTRONICS SECTION

ADDRESS

DATA

DATA ENTRY AND DISPLAY ASSEMBLY

OUT-OF-PLANE VELOCITY

X-POINTER INDICATOR

ALTITUDE AND ALTITUDE RATE INDICATORS

CLEAR

DATA READOUT

ENTER

HOLD

DATA ENTRY

ALTITUDE

ALTITUDE RATE

LEFT The Abort Guidance System was a critical back-up designed in to the LM when concern was expressed over the ability of the PGNCS to handle all operations, and to serve as back-up in the event of failure. Using less power, and with a simplified approach to maintaining position information and automated control of selected LM functions, the Abort Electronics Assembly was the heart of the AGS. *(Grumman)*

ABORT GUIDANCE SYSTEM

The Lunar Module had two separate loops for guidance, navigation and control, the primary being the PGNCS, the secondary being the Abort Guidance System (AGS). Both used the same hand controllers and both addressed attitude and translation capability through the same rocket thrusters and motors, but the AGS had a much simplified path and a secondary system of input and data processing.

The AGS consisted of an Abort Sensor Assembly (ASA), a Data Entry and Display Assembly (DEDA, pronounced 'deeda') and an Abort Electronics Assembly. The ASA consisted of three floated, pulse-rebalancing, single-degree-of-freedom, rate-integrating gyroscopes and three pendulous reference accelerometers housed in a beryllium block on the navigation base. It provided attitude and velocity changes in all three axes, with data pulses sent to the electronics assembly for processing.

The Abort Guidance Computer (AGC) was the brainchild of Eldon Hall at the Charles Stark Draper Laboratory. It weighed 70lb (31.7kg), measured 2ft x 1ft x 6in (61 x 30.5 x 15.2cm) and was made up from 5,600 three-input NOR gates in pairs of integrated circuits,

a very new technology for its day. The AGC was sealed in an aluminium magnesium box and the unit consumed 55W.

The DEDA was a very much simplified and less user-friendly version of the DSKY employed with the PGNCS, consisting of a basic keyboard with operational modes or data inserted via a three-digit 'word' with plus or minus value and a five-digit 'code' in octal characters. Readouts partially reversed the process, with a three-digit address of the desired 'word' followed by depression of a button to get the information. The DEDA was 7.3 by 6.6in (18 by 17cm) in size with a depth of 5.6in (14cm) and a weight of 8.4lb (3.8kg).

Information from the DEDA was transferred to the electronics assembly, a high-speed general-purpose computer with 18-bit instruction words consisting of a five-bit order code, an index bit and an 11-bit operand address. The memory was a ferrite-core stack with a capacity of 4,096 instruction words divided into two equal sections. The temporary memory stored replaceable instructions and data while the permanent memory stored instructions and constants that could be changed during the mission.

The ASA was a critical piece of equipment used in conjunction with the AGS (Abort Guidance System), the back-up, stripped-down, guidance and navigation equipment designed into the LM in case of a failure to the PGNCS system. The AGS would be used to control the upcoming course correction burn at MCC-5. The ASA was built to be maintained at a temperature of 120°F (48.9°C). NASA needed to know if it could be chilled down to very low temperatures, powered back up and still work effectively. Specialist teams at the Manned Space Center conferred with its maker – Hamilton Standard's Farmington, Connecticut, facility – and tests immediately began which indicated that it could. But there were no guarantees.

Of equal uncertainty was the condition of the motor switches in the Command Module that tie the battery buses to the main bus. If they were closed the electrical system in *Odyssey* could be controlled through the circuit breakers. North American were asked to quickly analyse the possibility of a problem with the switches due to low temperatures, and at 92:52 GET both EECOM and SPAN (Spacecraft Analysis, co-located with Mission Control) concurred that Swigert should be asked to do a check. SPAN supported the interface between Mission Control and the host of contractors and subcontractors and processed requests for support, a lot of which had been done in the past 34 hours.

When the modified checklist was read up to Swigert he closed the motor switches and checked both buses, reporting 32.3V on main A and 37.0V on main B with zero amps, as they should be. At 94:19 GET EECOM noted those satisfactory numbers on his log sheet, after which the circuit breakers were pulled and the switches left as they were – closed.

As the partial pressure of carbon dioxide built up in *Aquarius*, Mission Control watched until the level reached 7.5mmHg and then, at 93:53 GET, Fred Haise switched over to the 'mailbox' system, with one, hooked to Jim Lovell's (CDR) hose, up in the tunnel and the other, from Fred Haise's (LMP) hose, attached with duct tape to the ECS system on the right side of the LM.

Almost immediately the CO_2 level began to fall. In this configuration the LMP's red hose was on the ECS package with the LMP's blue hose against the right window, and the CDR's red hose was in the tunnel and the blue hose blowing up in to *Odyssey*. Within 15 minutes the level was down to 0.5mmHg, eventually falling to 0.1mmHg, which it would reach within 90 minutes. As to the crew's state of rest, Lovell had managed three hours on the last session, Swigert and Haise reported six hours each.

At 94:37 GET Capcom Kerwin voiced up a procedure to Fred Haise regarding the Descent Stage water tank: 'It's no big problem but our LM people say that the bottom of the Descent Stage is probably cooling off, and we just want to verify that the descent propellant tank will be OK. Right now it looks as though it won't freeze until several hours after its empty, but we want to have you read these temperatures out to us so we can see how good our predictions are. The procedure is, on panel 16, close the Propellant Display/Engine Override Logic circuit breaker...' and a host of specific instructions followed.

Panel 16 is located to the right of the LM pilot's window at head level and it supported four rows of circuit breakers. Haise read out the two fuel tank temperatures as 66.5°F (19.2°C) and 66°F (18.9°C), with the two oxidiser tanks showing 65°F (18.3°C) and 66°F. But further evidence of activity in the crippled *Odyssey* got Fred's attention: 'We've noticed some fresh particles floating around outside, so possibly the Service Module is starting to vent a little again.'

Planning for the upcoming trajectory correction addressed another consumables issue. Flight controllers were not only concerned about the oxygen, water, battery power and CO_2 scrubbers necessary for sustaining life, but also about the magnitude of velocity changes retained by the onboard propulsion systems. Given the current mass of the docked vehicles at 87,942lb (39,890kg), and based on the amount of helium pressurisation remaining, the descent propulsion system had approximately 800ft/sec capability. In other words, it could change the velocity (speed with direction) by that amount, equal to 545mph.

With each passing hour, as the velocity of Apollo 13 began to increase due to the Earth's gravitational pull, the magnitude of the course correction would increase. At a planned 7ft/sec (2.1m/sec) conducted at 105:00 GET that was not a problem with the amount of ΔV remaining

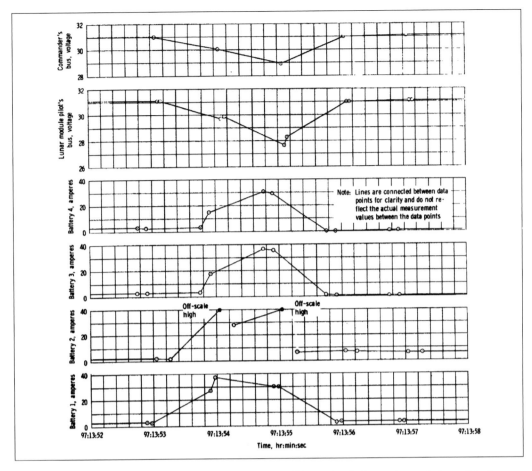

LEFT At around 97 hours elapsed time, one of the LM batteries burst its case due to a build-up of gas. This section of the telemetry chart from one of the consoles in Mission Control shows the spikes in amperage that occurred around that time. Readouts show what the two instrument displays in *Aquarius* were showing, in addition to the telemetry providing additional measurements down to Mission Control. (NASA)

in the DPS, but the helium disc was predicted to burst around 107:00 GET, after which the descent engine would be unusable.

The other propulsion capability was in the four RCS thrusters' quads, and at this stage of the flight there was a 28ft/sec ΔV capability – about 19.1mph, or four times the expected requirement for the trajectory correction. Theoretically the RCS could handle the velocity change but that would have consumed 25% of the remaining propellant in that system, and the need to conserve as much onboard propulsion as possible made early use of the descent engine, while it was still available, expedient.

Moreover, if left until 118:00 GET the magnitude of the change would have increased to a ΔV of 40ft/sec (12.2m/sec), well beyond the capability of the RCS. Also, there could still be additional trajectory correction burns needed further down the road, and with the DPS helium disc about to burst and vent the gas to space. That could again upset the fine-tuned trajectory and compromise re-entry alignment with the Earth's atmosphere unless corrected with RCS burns.

To survive that re-entry and stay within the physical deceleration limits of the human body, the structural limits of the Command Module and the capabilities of the heat shield, the entry angle to the local horizontal had to be within a tight band. This was known as the entry gamma. Steeper than -7.4° at entry interface and the CM would overstress the design and the heat load would cause it to burn up. Shallower than -5.25° and it would skip out like a stone on a pond and execute a highly elliptical path before once again returning to the vicinity of the Earth many hours, possibly even days, later. The optimum entry angle was -6.33° and this was the figure watched most keenly as ground stations tracked the incoming spacecraft.

There was another reason for selecting the early course correction. The SHe disc was expected to burst around 107:00 GET and although it had been designed as a non-propulsive event, with a T-pipe carrying opposing exits to cancel out thrust in any one direction, the offset centre of mass of the uniquely docked configuration was likely to influence the PTC roll and require attention, which could be avoided if the two events were relatively close together and handled at approximately the same time. Time management = efficient mission management!

'A little thump...'

Concerns once focused on keeping the crew alive were now switched to ensuring they had a healthy Command Module for re-entry. The mood in Mission Control was noticeably more upbeat than it had been the day before. However, many questions remained about the state of several vital systems within *Odyssey*, chilled down far below design temperatures and in some cases below the levels ever tested in development. The temperature in *Aquarius* stabilised at 52°F (11.1°C). In *Odyssey*, where the crew slept in cycles, it was 38°F (3.3°C).

For the crew it was not so much the low temperature that became difficult as the fact that it remained at that level for nearly three days. The medical personnel were getting very concerned, and as the hours wore on Fred Haise seemed to get colder than Lovell or

BELOW Jim Lovell tries to get a nap in the left forward crew station. *(NASA)*

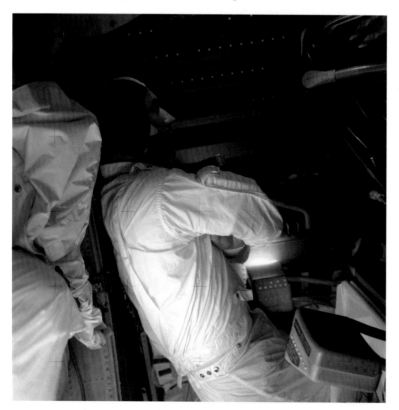

Swigert; then he started almost uncontrollable shivering and began to show signs of an infection. Pressure on the EECOM console sought ways to find a little power to warm things up a bit. There were none.

On a different shift cycle to the flight control teams, Vance Brand relieved Joe Kerwin about 95:30 GET and continued to talk with Fred Haise about particles continuing to vent, the PTC wobble getting worse, and switch configurations for checking the condition of the electrical buses in *Odyssey*. With Lovell and Swigert resting there was little conversation but still streams of particles were clearly visible. At 97:14 GET Fred Haise called down to Vance Brand: 'I just heard a little thump, sounded like down in the Descent Stage, and I saw a new shower of snowflakes come up that looked like they were emitted from down that way.'

What Haise saw emanating from the vicinity of sector 4 corresponded to current transients across all four Descent Stage batteries, with erratic traces flickering across EECOM's screen for two seconds beginning at 97:13:53 GET – another electrical malfunction that momentarily caused a heart-stopping expectation of something more serious than it turned out to be. Corresponding drops in DC bus voltage and a momentary blip in the water glycol pressure differential for the heat transport system indicated that the pump had slowed – briefly.

What had been observed on the LM pilot's side of the DC electrical system affected batteries 1 and 2 and immediately triggered a call from Bob Heselmeyer to a chain that landed at Don Arabian's desk in the Mission Evaluation Room. From there it brought in Grumman and Eagle Picher, the battery contractor. The reason for the thump sound and the electrical spikes was soon resolved into an annoyance but not a problem. After the surges battery 2 stabilised and, because they are operated in parallel and not in series, all four batteries balanced the current demand, now running at a super-efficient 11.6A.

The most likely cause was potassium hydroxide electrolyte leaking from one or more of the 20 cells in each battery bridging the high-voltage or low-voltage terminal to the battery case. This would result in water electrolysis and ignition of the trapped hydrogen and oxygen

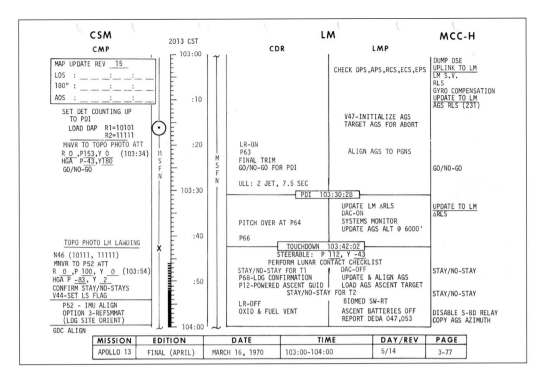

LEFT As they cruised home, getting colder by the hour in their powered-down 'lifeboat', thoughts could only fleetingly rest on what should have been happening at 103 hours into their planned mission. This Flight Plan page shows events associated with the planned Moon landing that never took place. (David Baker)

generated, producing a miniature explosion that would rupture the battery lid and produce the cloud of snowflakes that Haise saw through his window. Debonded potting allowing electrolyte to leak was a problem with these batteries, which could leach through the Teflon retention screens ostensibly fitted to prevent leaks.

Coming up toward the end of Glynn Lunney's shift, EECOM advised that the hydrogen tanks in the damaged Service Module should be venting. Without any power going into the electrical production system in *Odyssey*, the liquid hydrogen was sitting in the tanks unused, and that was calculated to have reached the vent level. The crew had been seeing venting and discharges for some time and there was a difficulty knowing exactly what was coming from where. Only a few select systems were powered up so telemetry was of little help.

The Gold Team came on duty at around 98:00 GET (3:13pm Wednesday 15 April) and Jerry Griffin organised an early Command Module power-on to get telemetry from the oxygen 'repress' packs, three 1lb (0.45kg) oxygen bottles installed to rapidly repressurise the Command Module in the event of depressurisation. When power-up came two hours later, reassuringly they were found to be at 820lb/in² (5,654kPa), as expected.

Within 30 minutes, having had four or five hours of sleep, Lovell and Swigert were up and getting ready to prepare for the course correction burn. Before that, a lot of instructions were read up to Jim Lovell for setting the attitude configuration. Then, at 99:51:09 GET, battery 2 played up again when the crew got a master alarm and a malfunction light. The crew were asked to turn the battery off while Houston looked at the telemetry. When switched on again the problem recurred and the battery was taken off line.

At 101:06 GET Jack Swigert began copying down a long list of procedures, switch positions and operations for partially powering up the Command Module using power from *Aquarius* so as to run a telemetry check on critical systems. From his couch in *Odyssey*, 30 minutes later Swigert began to return electrical power to the CM from battery B for a partial power-up. What the flight controllers saw was encouraging. Voltages, pressures and readings appeared as expected, and in some systems that could provide such information temperatures were warmer than expected. Although power-up on the CM lasted less than ten minutes, Mission Control got 7min 37sec of telemetry.

At 103:52 GET, TELMU noted that the temperature in *Aquarius* was being chilled below the level it needed to be. He suggested placing the suit temperature rheostat in the

full cold position. This reduced the flow of glycol, which normally is warm and heats the oxygen to the suit heat exchanger. Because the ambient temperature was so low, the glycol was having the reverse effect and chilling down oxygen coming out of the tanks. This action reduced the flow rate and increased the cabin temperature by a slight amount.

Preparations for the course correction burn went so well that the crew were ahead of their timeline and elected, with concurrence from Mission Control, to do the manoeuvre early. It was not dependent on a specific time and could be done as soon as they were ready. Because the MMC-5 burn was not critical on attitude and had a +/-1° deadband it was not necessary to use the power-hungry PGNCS. The Abort Guidance System would do the job instead, which would help conserve electrical power and cooling water too.

The crew could obtain the correct burn attitude by turning the docked spacecraft to position the COAS on the centre of the Earth, aligning the illuminated portion at the top of the reticle. The azimuth could be obtained by sighting the cusps of the sunset terminator on the Earth on the Y-axis reticle. This would place the X-axis perpendicular to the Earth's terminator for a retrograde manoeuvre conducted perpendicular to the flight path, and so steepen the entry angle. Pitch was obtained by placing the Sun at the top centre of the AOT.

In this alignment the AGS was used for a body-axis attitude hold, with manual use of the attitude needles to hold the alignment. The entire operation was unique to Apollo 13, another procedure developed in the mission simulators back on Earth but carried out by the crew. From their respective crew stations in *Aquarius*, Fred Haise controlled pitch and Jim Lovell handled roll using their hand controllers. Yaw was controlled automatically by the AGS. Jack Swigert would call the engine start and stop times. The technique replicated a procedure well known to Jim Lovell and developed for the flight of Apollo 8. It needed some adjustment in application but it was equally adaptable to operation from the LM.

The manoeuvre was to begin with an ullage burn using four RCS jets for ten seconds. The burn time was to be 15 seconds on a minimum throttle setting of 10%, which actually meant 12.6%. Shutdown would be commanded manually at the calculated burn time minus one second. This was so as not to do an overburn, which could be caused by sluggish accelerometers, or PIPAs (Pulsed Integrating Pendulous Accelerometers), in the Abort Sensor Assembly of the AGS. This had been powered down and cold-soaked for two days. Mission Control had highly accurate burn duration calculations, and the ASA could be sluggish in response to an onboard computer-controlled instruction.

Another cause for concern was that in the event of an overburn, residuals would have been nulled using the -X thrusters which could have impinged on the exterior of the Command Module, threatening the integrity of the heat shield. Another consequence of MMC-5 was the reduced pressure level in the supercritical helium tank, which delayed the time the burst disc was expected to rupture, so the spacecraft was put in a PTC based on the post-MCC-5 attitude.

The MCC-5 burn was made at 105:18:28 GET (10:31pm in Houston) with Apollo 13 now 175,179 miles (281,863km) from Earth at a velocity of 3,039mph (4,890kph). Some 52lb (23.6kg) of propellant had been consumed and *Aquarius* now had a mass of 25,233lb (11,446kg). Refined now, the trajectory parameters would be watched over the next several hours to determine whether Apollo 13 would need a further course correction burn. Within a couple of hours GUIDO was showing an entry gamma of -5.99°, a little high. *Aquarius* would have to last out a further 36 hours until it was cut loose around 141:30 GET.

As for the consumables, calculations showed that despite raising power consumption to 45A for the burn phase, averaging 23A for the power-up phase, with use dropping quickly back to 12A, there would be sufficient battery power to last out the predicted entry time. Water consumption had increased to 3.1lb/hr during the manoeuvre but quickly settled back to 2.6lb/hr, predicted to last through 160:00 GET. Oxygen was plentiful, predicted to last beyond 270:00 GET, and there was ample life in *Odyssey*'s LiOH canisters through the jury-rigged 'mailboxes' feeding through *Aquarius*' ECS.

Yet another day...

At 106:00 GET the Maroon Team came on shift in the MOCR. It was time too for Swigert and Haise to get some sleep, leaving Lovell on watch. With yaw controlled by the AGS, the PTC barbecue mode was set up by rolling 90°, placing the Earth's terminator parallel to the X-axis of the COAS, the rates nulled as far as possible in pitch and roll with the Thrust Translation Controller Assembly (TTCA). Accuracy in these axes was not possible due to fluctuations and inaccuracies in the needle indicators in *Aquarius*. Twelve small thruster pulses were used to start the PTC control mode, but because the rates could be completely nulled some pitch-roll coupling occurred.

While confidence was high in Mission Control, conditions in *Odyssey* were getting worse, the crew having to migrate back into *Aquarius* or the docking tunnel for sleep, sharing space with the crew member on watch. As early as 73:00 GET the windows in the Command Module had become opaque with water droplets, the temperatures stabilised only a few degrees above freezing.

Within a few hours after the MCC-5 burn, the walls of *Odyssey* began to grow a thin film of moisture and a clammy feeling to the atmosphere could be felt in some corners. And now water was forming on the windows in *Aquarius* as well. Generally the air was not so moist as to be uncomfortable to the crew, but the cold was remorseless, with conditions much like those encountered on entering a walk-in refrigerator. Of concern was the condition of the switches, dials, meters and controls in *Odyssey* and whether they would work effectively, or whether the moisture and water droplets would cause an electrical short that could leave the crew's only haven powerless and inert.

But one more milestone was crossed at 107:00 GET. Should something catastrophic happen to the Descent Stage, which held the main bulk of the consumables, the life support functions of the Ascent Stage alone were now capable of lasting out the mission. At 108:54 GET (2:07am on Thursday 16 April) Jim Lovell responded to a call from Jack Lousma on the Capcom console asking him if he had spotted anything: 'Yes, Jack. I was just about ready to call you. Underneath quad 4, I noticed a lot of sparklies going out.'

The supercritical helium tank had finally let go, venting the remaining 27.5lb (12.5kg) of gas overboard. When the SHe disc burst the pressure had reached 1,937lb/in² (13,356kPa), but a surprise lay in the disturbance to the attitude of the spacecraft caused by this 'non-propulsive' vent. Because of an adverse pressure distribution in the two vent ports 180° apart, the two plumes had an included angle closer to 90°, and the net thrust resulted in a roll imparted to the spacecraft as it slowly bled empty in approximately ten minutes.

BELOW Given the problems with the supercritical helium tank before launch, it was inevitable the burst disc would pop open if the helium was not used during a lunar landing, the time for which had already passed. The burst disc was designed with stainless-steel Belleville washer springs which controlled the setting. During some tests with other tanks, the disc burst at 300lb/in² (2,068kPa) below the predicted level, this being due to corrosion caused by long-term exposure to moisture, and as a result some discs were refitted with Inconel springs. (NASA)

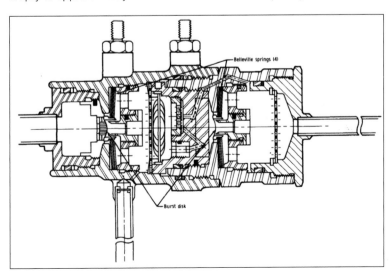

Within minutes of it starting, Swigert was on the microphone calling out the attitude changes. From a previous PTC roll of one revolution every 18 minutes, the venting changed roll direction and increased the rate to once every 3min 50sec. SPAN did an assessment and advised the flight activities officer (FAO) that the only limits were structural: no spin rate faster than one revolution every ten seconds.

Several hours earlier, not long after midnight on the morning of Thursday 16 April, Gene Kranz stood up in Room 210 and ordered his team out the door. Not since he pulled his White Team from the MOCR more than 40 hours before had there been any rest. Nobody had managed to sleep, catnaps being the only relief from tiredness held at bay by adrenalin and intense, unremitting work. The controllers were reluctant to leave, doubting the logic of letting go with the crew still in deep space. But the Tiger Team would revert to the White Team for duty in the MOCR during the transfer to the Command Module and re-entry, and they needed their rest.

Elsewhere, at 2:00am a meeting was held with the SPAN people, EECOM, TELMU and the back-up crew to finalise the entry procedures and timeline. A ten-hour meeting the previous day had struggled rigorously with various options for those events, drawing expertise from contractors and engineering personnel from across government and corporate facilities.

Aboard Apollo 13, at 110:01 GET, Jim Lovell called down with a progress report on the sleeping accommodation in *Aquarius*: 'Jack and Fred are both going to sleep. It's sort of humorous; Fred's sleeping place now is in the tunnel, upside down with his head resting on the ascent engine cover. Jack is on the floor of the LM with a restraint wrapped around his arm to keep him down there.' Soon Fred Haise came on line at the end of what he confirmed had been a good sleep, despite the odd position, and got ready to copy down a procedure for applying LM power to battery A in *Odyssey*.

The three CM batteries each held about 40Ah, but battery A was down about 20Ah and by drawing approximately 8A off the LM supply the crew would be assured three full batteries for re-entry. The three silver-zinc storage batteries were located in the lower equipment bay, each with 20 cells with potassium hydroxide and water as an electrolyte. The plastic cases were coated with fibreglass epoxy with any outgassing vented overboard.

Each battery was 6.8in x 5.7in (17.3cm x 14.5cm) in size and weighed 28lb (12.7kg), providing a high power-to-weight ratio. Open circuit voltage was 37.2V and a minimum of 17V could be maintained through to depletion. The functional purpose of the three batteries was to supply all the electrical needs of the Command Module after separating from the Service Module or, in the case of Apollo 13, the Lunar Module. Charging began at 112:11 GET (5:24am in Houston) and was expected to last approximately 15 hours.

Beginning at 113:15 GET, Fred Haise was asked to tape up a second pair of LiOH canisters to the existing 'mailboxes', and a precise procedure to do that was voiced up by Jack Lousma. CO_2 readings were 1.8mmHg, and with the first one strapped on by 113:43 and the second 13 minutes later the level dropped appreciably. Both Swigert and Haise were on watch but Jim Lovell was already an hour into his rest period, such as it was with temperatures now very cold in *Odyssey*.

BELOW The forward station of *Aquarius* with the Commander's 'eight ball' visible.
(Grumman)

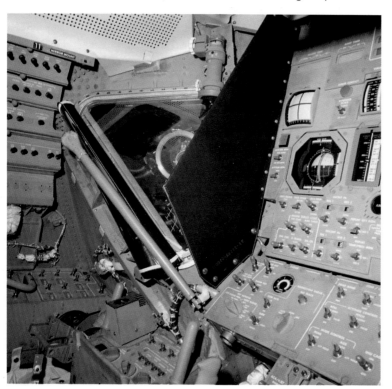

Glynn Lunney's Black Team had signed on at 113:00 GET with the entry gamma now hovering around -6.25°, the angle a little shallower than desired. FIDO was already looking at doing a tweak burn five hours prior to re-entry. The all-important number was defined as the vacuum perigee, the low point at which the spacecraft would pass across the Earth, assuming no atmosphere was present. To reach the surface of the Earth the perigee point had to be within the atmosphere so that the kinetic energy would be discharged as heat through friction with the air, slowing the spacecraft.

By extracting the energy that would otherwise take it on a long elliptical trajectory past Earth, the spacecraft would be drawn further into the atmosphere eventually, converting a near-horizontal flight path into a vertical descent on parachutes. With an entry gamma of -6.25° the Command Module would reach a vacuum perigee of 26.5 miles (42.6km) at a velocity of 36,431ft/sec (11,104m/sec) or 24,839mph (39,966kph), some 221ft/sec (67.36m/sec) or 150.7mph (242.4kph) faster than entry interface at an altitude of 400,000ft (121,920m) or 75.76 miles (121.9km). Entry interface is defined as the point where the density of the atmosphere is first felt by a spacecraft and a deceleration of 0.05g begins. FIDO wanted to get the vacuum perigee down to 23 miles (37km).

One hour into the new shift, the regular consumables update was logged by the flight director. The situation was now stabilised and had been for some time. The gross characteristics of the docked vehicles were crucial to the upcoming events, both for the course correction and the procedures for dumping the Service Module and *Aquarius*. The total stack had a mass of 87,740lb (39,799kg), of which 58,728lb (26,639kg) was propellant.

Consoles showed 33.17lb (15.05kg) of oxygen remaining in the Descent Stage tank, 2.25lb (1.02kg) in Ascent tank No 1 and 2.67lb (1.21kg) in Ascent tank No 2. Water quantities showed 51.9lb (23.54kg) in the Descent Stage, 42.1lb (19.1kg) in Ascent tank No 1 and Ascent tank No 2. Cabin pressure was holding at 4.98lb/in² (34.34kPa). With some 8A being drawn for recharging CM battery A,

current usage was up around 17–18Ah, with a prediction for the LM batteries to last until 170 hours at that level. And this was at least one LM battery activity that did nothing to increase water use, because no additional systems were powered up to need cooling.

At 115:38 GET, with Haise still asleep, Lovell relieved Swigert who began his rest period although, since it was too cold to sleep in *Odyssey*, there was little opportunity for a proper rest. The men were already wearing two sets of underwear and Lovell and Haise were wearing their lunar boots. 'You can tell the guys in Crew Systems that lunar boots make great foot warmers,' said Lovell, bantering with Joe Kerwin, now at the Capcom console.

At 118:44 GET (11:57am Thursday 16 April) Houston informed Lovell, 'Incidentally, you've less than 24 hours to go.' At 119:44:33 GET Apollo 13 logged another milestone – the halfway distance between the Earth and the Moon, the spacecraft now 127,428 miles (205,032km) from each. The consumables were holding up well. Electrical power was drawing 17.9A, 974Ah remaining, water was being consumed at 2.5lb/hr with 111.4lb remaining, and oxygen was going down at 2.5lb/hr with 33.43lb on board. Cabin pressure was at 4.94lb/in² and the CO_2 level was a healthy 0.1mmHg. In Houston it was 12:57pm. With luck, in little more than 23 hours the crew would be floating in the warm waters of the Pacific Ocean.

ABOVE As Earth drew closer, thoughts of getting back alive focused on the preparations for powering up *Odyssey* and checking systems that had been exposed to temperatures and moisture levels for which they were never designed. *(NASA)*

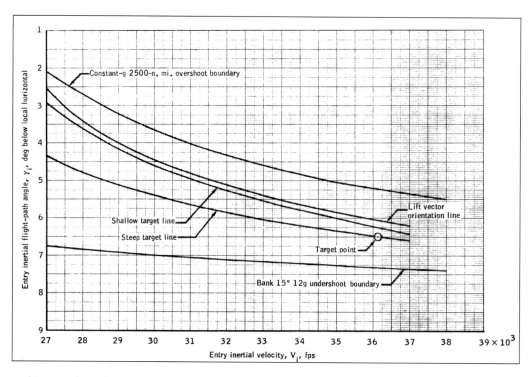

RIGHT Re-entry was becoming the dominant discussion during the final 24 hours of the mission. The entry gamma, or flight-path angle to the local horizontal, was critical – too steep and the spacecraft would burn up, and too shallow and it would skip back out of the atmosphere like a stone skimming across water. This plot shows the corridor through the trajectory along which the spacecraft had to thread a re-entry path, ideally an entry gamma of 6.5°. *(NASA-MPAD)*

At approximately 120:00 GET Gold Team relieved Black Team and Griffin got ready to do a press conference, now a regular chore after each shift handover, while Lunney settled in for a long briefing session with the crew. Two days of highly detailed and complex activity would involve a further powering up of the Command Module to check telemetry and get a tighter handle on the state of the spacecraft. And throughout the shift, there were detailed briefings on procedures for packing *Odyssey*, stowing unnecessary items in *Aquarius* and working out where everything was to go in this most unique of missions.

Although conducted in the back rooms at Mission Control, in SPAN, where the contractors were helping provide minute details on equipment and components, and in the Mission Evaluation Room, where Don Arabian was carefully working out what readings, instrument settings and telemetry trace graphs meant, the support work was vital to getting the crew through these final stages of the mission. In Room 210 the real architects of consumables budgeting and of devising ways to achieve it were, respectively, John Aaron and Arnie Aldrich. Backing them were brilliant minds working precise switch positions and power levels.

At 120:22 GET all three crew members

listened as Brand read up a general outline of the procedures to be followed prior to re-entry. The overall plan was to begin preparations for entry at EI-6.5 hours (6h 30min before entry interface), power up the LM at EI-6 hours and conduct a final course correction burn, known as MCC-7, at EI-5 hours. The burn would be conducted with the LM RCS thrusters and be no more than 2ft/sec (0.6m/sec), about 1.36mph (2.19kph), details to be refined later.

The Service Module would be jettisoned at EI-4.5 hours and the attitude of the docked LM/CM set so that with a simple pitch manoeuvre the crew would be in a position to take photographs of the damaged module as it drifted away. But there would be no other changes of attitude to fully optimise that sighting, for fear of upsetting the trajectory. Full Command Module power-up would begin at EI-2.5 hours, followed by closeout of *Aquarius* and manoeuvre to the separation attitude at EI-1.5 hours. Jettisoning the LM would come at EI-1 hour.

Immediately thereafter the CM would manoeuvre to entry attitude, with the crew doing a sextant star check on the night side of the Earth, taking down the optics and then receiving the final entry pad from Mission Control. With this they would initialise the Entry Monitor System (EMS), the integrated display situated on the top left of Panel 1 directly in front of the commander's seat. The EMS presents five separate displays for monitoring an automatic re-entry or for supporting a manually controlled re-entry. The threshold indicator monitors deceleration levels, velocity vectors, roll indication, corridor verification and flight-path range.

As originally planned, the Command Module would have entered the atmosphere in daylight and the crew were to have conducted a horizon check to confirm the guidance and navigation data. As it was, the CM would now enter on the night side of the Earth, but the geometry of the relative position of the Earth allowed the crew to do a Moonset check instead at about EI-3 minutes. After read-back Lovell had some questions for Mission Control on the general outline of procedures, including his recommendation that they should not clamber into bulky space suits for entry.

Mission Control also outlined the unique way the Service Module was to be jettisoned. Under normal mission circumstances the SM would continue to thrust away from the Command Module after separation, using electrical power from the fuel cells to control and operate the RCS thrusters. With a dead SM, there was no way of operating those thrusters. For Apollo 13 the separation would be conducted in what they called a 'push-pull' manoeuvre using the LM thrusters, first pushing forward 0.5ft/sec +X (0.15m/sec), releasing the module, then backing away with a 0.5ft/sec –X burn. By thus effectively imparting a 1ft/sec (0.3m/sec) separation velocity the two should be 16,200ft (4,938m) apart at re-entry, thereby lessening the risk of collision.

For several hours Fred Haise's health had been going downhill, his sweaty feverishness turning into something more ominous as time passed. The medical team was concerned lest a lack of fluids should weaken the crew's ability to excrete toxins, leaving the body susceptible to attack. All three crewmen were vulnerable, but Haise was deteriorating and suffering largely in silence, resisting the attentions of his fellow travellers and downplaying the symptoms of what was shaping up to be a nasty kidney infection.

The crew had been consuming a little over 6oz (177ml) of water per person per day. Water was at a premium and there was only one source, serving the cooling system, the crew's needs and, theoretically, the rehydration of their food. No water was used for dehydrated meals and cold food had to suffice for the more than three and a half days between the accident and splashdown. Moreover, with a desire not to vent fluids overboard for fear of causing a propulsive 'thrust' that might affect the trajectory, they were bagging urine and storing it wherever possible.

As the crew got ready to copy down the

LEFT **Most of the procedures coming up involving both** Odyssey **and** Aquarius **were tested over and over in the Apollo Mission Simulator at the Manned Spacecraft Center, where astronauts rehearsed procedures and modified the sequence of events, unique to this mission, that would only be voiced up to the crew when fully evaluated.** (NASA)

complex procedure for all the events that were to take place in the final hours of this mission, battery A in *Odyssey* achieved full charge and, at 126:03 GET, a boost charge on battery B began, terminated at 127:59 GET.

There had been almost no indication of any tension between Apollo 13 and Mission Control during the 70 hours since the accident, but lack of sleep, exhaustion and frustration over waiting for the detailed checklist to be delivered to the MOCR began to tell. At 126:07 GET, Lovell called down: 'Vance, you've got to realise that we've got to establish a work-rest cycle up here, so we just can't wait around here to just read procedures all the time up to the burn. We've got to get them up here, look at them, and then we've got to get the people to sleep. So take that into consideration when you get ready to send up the pads.'

And from Mission Control: 'I know Jim. We're very conscious of that. We – we should be ready to go in about five minutes. That's all I can say. Stand by.' Five minutes later Brand was back and started the read-up and Swigert began to copy, within minutes handing over to Ken Mattingly, who had worked the procedures in the simulator. Then the read-up switched

to Fred Haise for all the LM deactivation and separation procedures. This lengthy series of instructions had to be copied down by hand aboard *Aquarius* and read back for confirmation. The read-up began at 126:15 GET.

During this period, around 126:45 GET, with Capcom talking procedures up to the crew, Milt Windler's Maroon Team came on shift, exchanging updates and conferring on upcoming events. Finally, about an hour later, Griffin was off to go talk to the press with public affairs officer Jack Riley and FIDO Bill Stoval at his side.

Aboard Apollo 13 Fred Haise stirred, and was up talking to Mission Control by 128:00 GET, and 16 minutes later Capcom Vance Brand shifted the read-up from Ken Mattingly's Command Module procedures to LM power-up checklists and had a welcome recommendation for Jim Lovell: 'We've still got a LM update but we strongly recommend that you and Jack work as hard as you can sleeping for about the next five hours. Fred can handle it,' adding: 'You're free to drink all the Descent Stage water that you want.'

With little more than a day to go before splashdown, another critical decision had to

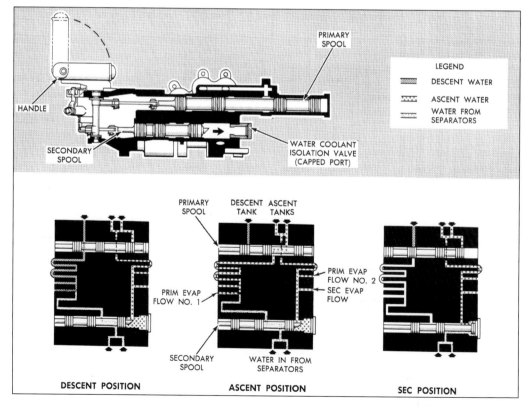

RIGHT A decision over when to switch from Descent Stage water tanks to those in the Ascent Stage was critical to maximising the water remaining for cooling electronic equipment that would soon be powered up for critical hours before separation of the spacecraft and re-entry. This selector valve is manually operated, and was used by the crew to make the switch after lengthy calculations by Mission Control and its contractors. *(Grumman)*

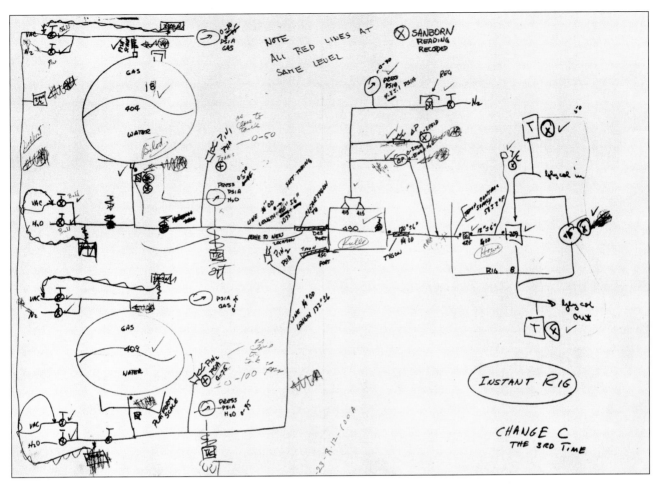

be made earlier that day, which only specialist engineers who had designed and built the LM ECS could address. With the water tank in the Descent Stage running dry, exactly when should the supply be switched by the crew to the two water tanks in the Ascent Stage? Should it be left to run dry and then switch? Or would that cause a problem in the sublimator? Ideally every last drop of water from the Descent tank should be used before switching.

Hamilton Standard's James Morancey came up with the answer when he devised a test rig to find out, and had drawings ready by 9:00am local time (8:00am in Houston) Thursday morning. With a complete duplicate set-up the same as the configuration aboard *Aquarius*, but with pressure sensors and monitoring meters, he had the test rig built by late afternoon and confirmation of the results to Mission Control by 8:00pm Houston time. It showed a preference for switching before the water ran dry, thus preventing the sublimator freezing over. At 9:33pm (128:20 GET) telemetry showed an

erratic pressure sensor and the crew switched to the Ascent water tanks, with a calculated quantity of about 5% remaining in the Descent Stage tank.

It was 129:33 GET when Haise completed the read-back to Vance Brand – a tiring 3 hours 20 minutes of dictation covering both *Odyssey* and *Aquarius*! Before signing off six minutes later to give him some respite, Mission Control updated the entry gamma, which had drifted now to -6.01°, indicating the need for a 2.81ft/sec (0.86m/sec) manoeuvre at MCC-7 – just eight hours away. Respite too for Vance Brand, who handed over the Capcom console to Jack Lousma.

For clarity on the readouts for procedures and timelines, the crew had switched on the LM power amplifier to boost the RF signal, but that was now switched back off, reducing the transmitter power from 20W to about 1W and saving some battery power. With Lovell and Swigert trying to get up to five hours' rest, and Fred Haise trying to catnap, there was no need for high fidelity voice.

ABOVE With water the critical consumable, Hamilton Standard's James Morancey sketched this plan of how to build a water-management test rig to work out the optimum path for conserving this valuable commodity. Without water, the cooling system would cease to function, the electrical systems would overheat and stop working, and *Aquarius* would shut down. *(David Baker)*

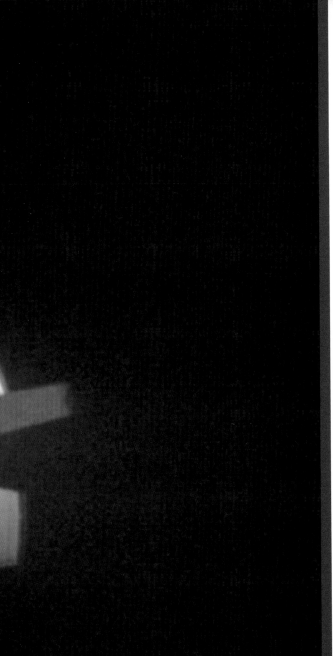

Chapter Seven

Home run

───⟨●⟩───────────

Tension mounted as preparations for re-entry were worked out on slide rules and rehearsed in simulators. Would *Odyssey* come alive when powered up? Or had it already succumbed to the cold and damp conditions, a lifeless hulk unable to support life? In Mission Control, nerves frayed to breaking point were held tight under icy control as flight controllers nursed the crew through the final desperate hours.

OPPOSITE As the Service Module floated away from the Command Module, the astronauts could see the damage done by pressure splitting apart the No 2 oxygen tank in sector 4, ripping off the panel. The tank did not explode, as remnants of it were still visible. It had split due to over-pressurisation caused by a fire or significant heat source rapidly turning the cryogenic oxygen into a gas. *(NASA)*

At 132:28 GET (1:28pm Thursday 16 April) Lousma called up to alert Fred Haise to some minor amendments to the LM activation checklist, and within a couple of minutes was speaking to Ken Mattingly about some changes to the CM list. Jim Lovell too was awake and chipping in with some points for clarification of procedures. This was the final wake-up before splashdown and with less than ten hours to go there was a lot to do.

During their most recent sleep Mission Control had been busy tracking and had fixed the entry gamma at -6.03°, still drifting unexpectedly. They had calculated the lift over drag (L/D) of the Command Module packed with items for return, and found it to be an acceptable 0.29, off only marginally from the nominal 0.31. With consumables looking in good shape there was some discussion about powering up *Aquarius* a little early and raising the temperature to improve crew morale.

Having had little more than three hours' rest, confirming that it was almost impossible to sleep in *Odyssey*, there was some need to balance available power and gradually warm up the spacecraft. But there was concern about the condition of the windows, the crew remarking again that they appeared to be frosted over, and Lovell was concerned that hitting them with a thermal load through the electric heaters could crack the panes.

At 133 hours there was a 759Ah battery capacity in *Aquarius*, with a predicted life of 195:00 GET at the present drawdown of 12A. Finally, there seemed to be a modest amount of power to spare. At 133:22 GET Lousma called Lovell: 'OK skipper, we figured out a way for you to keep warm. We decided to start powering you up now and what we want you to do is...', and there followed a switch list procedure for starting that process.

That would begin taking large amounts of power out of the batteries, but this action was only just workable because increasing the electrical drawdown would not only reduce the power remaining but would increase water consumption too. Within two hours, with power drawdown now up at 41Ah, battery life was tumbling and would last to 151 hours. Water demand had already increased to 4.8lb/hr (2.04kg/hr), cutting predicted life from an elapsed time of 156 hours to 148 hours. Within a few hours it would increase further to 5.4lb/hr (2.45kg/hr) as battery demand fluctuated according to the use of various systems, reaching a maximum 45.9Ah at one point, far higher than at any other time in LM operations as, in effect, *Aquarius* ran two spacecraft.

Balancing water consumption and assigning use levels to different factors was a particularly fine art. During much of the flight since the accident, when the consumption rate was around 2.95lb/hr (1.34kg/hr), only 1.35lb/hr (0.61kg/hr) was attributable to electrical loads.

About 1.17lb/hr (0.53kg/hr) was due to crew metabolic activity and LiOH loads based on an average 1,220Btu/hr (1,286kJ/hr) and oxygen uptake levels. A further 0.43lb/hr (0.2kg/hr) was due to structural heat loads while 0.4–0.45lb/hr (0.18–0.2kg/hr) was attributable to waters used by the LM separators in the ECS.

Between 134:45 GET and 135:02 GET, marks were taken on the Sun and the Moon because the star field was saturated with 'sparklies' coasting along with the spacecraft. This P52 LM alignment resulted in large torque angles that were aligned out at 135:04:35. Because this initial orientation was not the one planned, a P52 option 1 command was set up through the Lunar Guidance Computer (LGC) to realign to the desired values.

The LGC uses an algorithm and standard ephemeris data for computing out pointing errors, including Sun/Moon sightings as well as star sightings. For star angle differences the installed algorithm is used and resulting vectors are computed, with the arc cosine of the value giving the best estimate from the guidance computer of the angle between sighting vectors.

Right about this time there were believed to be sufficient consumables to power up the PGNCS, which was brought on line at 135:10 GET with consideration of using this and a P41 instead of the AGS for MCC-7. But the detailed instructions given to the crew had them set up the systems for AGS using the TTCA for attitude control and the PGNCS in only a monitoring role. Nevertheless, this change of plan anticipated using the PGNCS in its traditional role of prime backed-up by the monitoring role of the AGS.

Just prior to the PGNCS alignment, however, the AGS body axis alignment at a known inertial orientation with respect to the Earth was performed at 137:06 GET, and produced excessive use of RCS propellant. The LM was course aligned to this attitude, with the equivalent REFSMMAT stored in the Lunar Guidance Computer, but when placed in Auto – whereby the LGC would fine-tune the alignment – the Flight Director Attitude Indicator (FDAI) needles failed to zero to confirm proper convergence. Fearing there was something wrong with the PGNCS, at 137:15 GET the operating plan reverted to the AGS as prime, since it had worked well the previous time.

Ignition of the four RCS thrusters for MCC-7 came at 137:39:48.39 GET with Apollo 13 at a distance from Earth of 43,510.1 miles (70,007.8km), barrelling in toward the home planet at 6,982.4mph (11,090kph). *Aquarius* now weighed 25,181lb (11,422kg) and the CSM weighed 62,468lb (28,335kg). The burn lasted 21.5sec and produced a ΔV of 2.91ft/sec (0.89m/sec).

The purpose of MCC-7 had been to change

BELOW Flight controllers gather around Flight Director Glynn Lunney (seated with headphones) in the MOCR during the final hours before re-entry. *(NASA)*

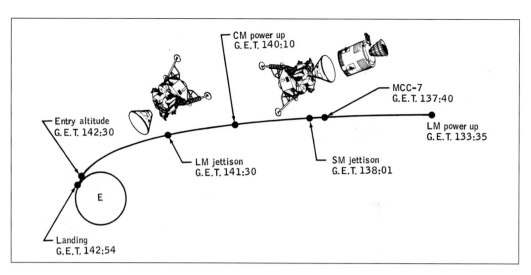

RIGHT The events scheduled for the final hours in the flight focused on powering up the Command Module, separating from the wrecked Service Module, and releasing *Aquarius* for destructive re-entry over the Pacific Ocean. *(NASA)*

BOTTOM The Command Module had a lift vector with L/D (lift-over-drag) varying according to angle of attack (α) and the velocity, as shown in this schematic. Because the CM had a higher L/D than previous spacecraft, it had the ability to re-designate uprange or downrange to compensate for errors at entry interface. The rounded edge of the base of the conical module was necessitated by the requirement to alter pitch angle during descent from hyperbolic trajectories, and to maintain the much hotter hypersonic shock wave some distance from the surface. With an acute, non-rounded angle between the base and the sidewall, as with Mercury and Gemini spacecraft, the hot shock wave, which must remain detached, could attach itself to the surface and burn through the heat shield. *(North American Rockwell)*

the EI gamma from -5.9° to -6.3° and this was achieved, with a vacuum perigee of 23.6 miles (38km). But in the final five hours to re-entry the gamma prediction continued to drift to -6.2°, although this did not change the decision to do re-entry as lift-vector 'up'. Only after the mission would post-flight analysis (to which your writer was committed) demonstrate that the gentle 'steaming' of the LM's water sublimator had produced a very slight delta velocity to ease the flight path away from the predicted value.

After the docked vehicles were manoeuvred to separation attitude by the thrusters on *Aquarius* at 137:57 GET, the guidance platform heaters in *Odyssey* were turned on

and the Command Module RCS thrusters pressurised and hot-fired at 137:57 GET to test their condition. Adopting an abort guidance reference, the LM went to attitude-hold mode using the AGS, X-axis translation being monitored on the displays. Following systems checks on the RCS thrusters, Lovell performed a +X translation of 0.5ft/sec (0.15m/sec), effectively 'pushing' the Service Module away.

Physical separation came at 138:01:48 GET with an audible firing and physical shock of the pyrotechnics to sever the connections, followed immediately by a -X translation of the same magnitude, backing the docked vehicles away from the jettisoned Service Module. The

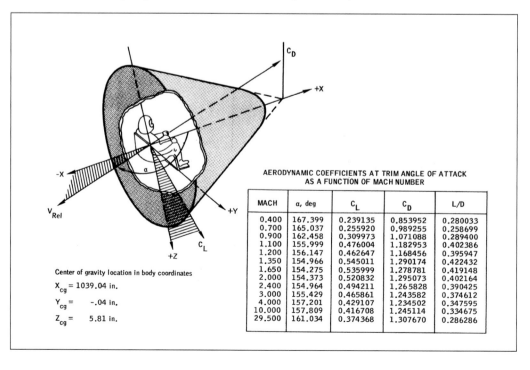

AERODYNAMIC COEFFICIENTS AT TRIM ANGLE OF ATTACK AS A FUNCTION OF MACH NUMBER

MACH	α, deg	c_L	c_D	L/D
0.400	167.399	0.239135	0.853952	0.280033
0.700	165.037	0.255920	0.989255	0.258699
0.900	162.458	0.309973	1.071088	0.289400
1.100	155.999	0.476004	1.182953	0.402386
1.200	156.147	0.462647	1.168456	0.395947
1.350	154.966	0.545011	1.290174	0.422432
1.650	154.275	0.535999	1.278781	0.419148
2.000	154.373	0.520832	1.295073	0.402164
2.400	154.964	0.494211	1.265828	0.390425
3.000	155.429	0.465861	1.243582	0.374612
4.000	157.201	0.429107	1.234502	0.347595
10.000	157.809	0.416708	1.245114	0.334675
29.500	161.034	0.374368	1.307670	0.286286

Center of gravity location in body coordinates

$X_{cg} = 1039.04$ in.

$Y_{cg} = -.04$ in.

$Z_{cg} = 5.81$ in.

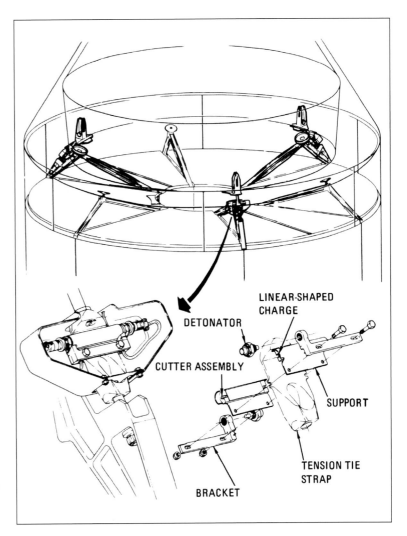

two modules had been held together by three tension ties 120° apart protruding from the base heat shield on the Command Module. These were stainless steel straps 4in (10.2cm) long by 2.5in (6.3cm) wide. The CM reste down on six compression pads, three 4in in diameter and three 6in (15.2cm) in diameter, raised out from the six radial beams dividing the six SM sectors.

The CM and the SM were also connected by a cluster of umbilical lines for electrical power and fluids including gaseous oxygen, water and coolant carried between the two during normal operation. These lines and pipes were clustered into a bulbous enclosure 40in (101.6cm) long and 18in (45.7cm) wide situated on the +Z (back) side of the Command Module, below the optics and on the opposite side to the main entry hatch. They fed down into Service Module sector 4 directly above the three fuel cells.

To sever the links between the SM and the CM, valves on either side of the separation plane cut off fluid flow and electrical circuits were deadfaced. Driven by redundant detonators, a guillotine with two sets of stainless steel cutters severed the lines, and the tension ties holding the two modules together were severed by linear shaped charges just 0.1sec after the wires were deadfaced. The fairing enclosing the umbilical lines and tubing swung outward and was retained by the Service Module.

On missions up to and including Apollo 11, 2sec after separation a sequencer in the SM fired the roll thrusters for 5.5sec and the +X thrusters were fired to depletion. By so doing the module supposedly moved away from the CM and skipped off the top of the atmosphere into a high-apogee orbit only later returning

RIGHT The umbilicals carrying wiring and plumbing lines from the Service Module to the Command Module are carried in a shielded enclosure attached to the SM but designed to swing back and away from the CM at separation. *(North American Rockwell)*

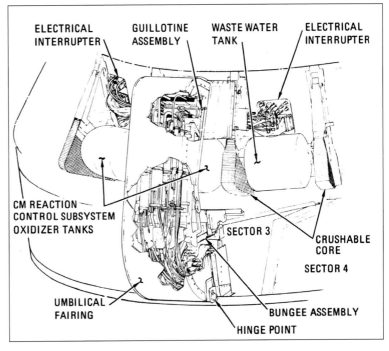

to destructive re-entry. On Apollo 11 the crew noticed their SM tumbling as it overtook them, its -X thrusters still firing, before breaking up in the atmosphere near the Command Module. For Apollo 13 the sequence had been changed so that the roll jets fired for 2sec and the -X thrusters for 25sec. Without power in the SM this was impossible, hence the unique 'push-pull' method employed achieving a zero net ΔV.

Service Module separation occurred at a distance from Earth of 41,076.7 miles (66,092km) and a speed of 7,094.7mph (11,415.4kph). The physical violence of the events caused the LM/CM to pitch down 10° and compromised the opportunity to sight the SM through window No 5 in *Odyssey*, where Swigert was positioned to get some photographs. Instead, using thrusters on the LM, Lovell pitched back up to sight the SM through the overhead docking window in *Aquarius*, where he and Haise were at their respective stations.

It first came into view at a distance of about 80ft (24.4m) and brought a gasp from the crew as they saw for the first time the state of the wrecked sector 4 containing fuel cells and what was left of the cryogenic tanks, torn Mylar insulation hanging out the side. Audibly stunned by what he saw, Jim Lovell exclaimed just three minutes after separation, 'And there's one whole side of that spacecraft missing...Right by the high-gain antenna, the whole panel is blown out, almost from the base to the engine,' adding minutes later, 'And Joe, looks like a lot of debris is just hanging out the side...'

The panel covering sector 4 was a composite of an outer (0.02in/0.05cm) face sheet fabricated from 2024-T81 aluminium and an inner (0.01in/0.025cm) face sheet from 7178-T6 aluminium, with an inner honeycomb core of 0.992in (2.52cm) thickness. The outer face sheet had a width of 67.04in (170.28cm) and a height of 151.94in (385.93cm), with a total panel mass of 98.8lb (44.82kg). Later tests demonstrated that the panel would blow out if pressure on the interior built up to 25lb/in^2 (172.4kPa).

At separation, 40,529.7lb (18,384.3kg) of propellants remained in the four Service Propulsion System tanks, some 99.5% of the loaded total, the modest difference having been consumed during the one hybrid transfer burn more than 107 hours earlier. The four RCS thruster quads had used 286lb (129.7kg) of the 1,342.8lb (609.1kg) of hypergolic propellants loaded for launch. Their last use had been just prior to total CSM power down at 58:40 GET. At that time in the mission the pre-planned use was a predicted 178lb (80.7kg). The excess had been consumed during 2hr 45min of repeated efforts to offset the destabilising effect of venting oxygen from the two cryogenic tanks.

As for the cryogenic tanks, at the time the CSM systems were shut down within two hours of the accident, of the 58.2lb (26.4kg) of liquid hydrogen loaded for launch 44.2lb (20.05kg) remained in the two tanks at the time of the incident, with an unknown quantity retained through to separation. Similarly, of the total 654lb (296.6kg) of oxygen in the two tanks, 157lb (71.2kg) remained when the contents of tank No 2 were lost almost immediately and tank No 1 evacuated itself within two hours.

With photo activity terminated at 138:14 GET the crew focused on getting the attitude set in preparation for Command Module inertial platform alignment. Because re-entry would be conducted under auto-guided control the IMU had to know the reference matrix. Determinations on how to achieve this occupied much of time spent by the White Team on working re-entry procedures, and the selected mode was known as the 'docked alignment' transfer option. Following an initial course alignment the platform in *Aquarius* would be fine-tuned to a greater level of accuracy by optical sightings with the sextant in *Odyssey*.

Jettison attitude was achieved at 138.20 GET, the Command Module Computer (CMC) was powered up in *Odyssey* and a fictitious DAP load for CM weight of 9,050lb (4,105kg) was placed in it. All this activity in the CM was powered by *Aquarius* alone. There had been some concern in the MOCR about the consumption of LM RCS propellant, a 13% uncertainty in the telemetry readings calling for a wide margin for possible measurement errors. Therefore with a 32% reading, the amount remaining that could be relied upon was no more than 19%. However, Swigert did manage to attempt some star sightings, although reflections off the LM sublimator and an RCS quad prevented useable acquisition through the scanning telescope, despite Lovell manoeuvring the combination by up to 20° and using more RCS.

But overall the consumables were right on the predicted line, and there was an encroaching sense that the worst was behind them, a certain levity in communications not heard for several days when Capcom Joe Kerwin tossed a career warning to Lovell: 'Well James, if you can't take any better care of a spacecraft than that, we might not give you another one!' The comment was ignored, Jim Lovell having already promised his family before launch that this would be his final mission.

At 139:15 GET LM battery No 3 was taken off line, essentially depleted with a reading of 1A, and battery No 4 was showing zero by 139:34. From then on events moved with a degree of controlled haste. Power from *Aquarius* to *Odyssey* was removed at 140:09:55 GET, some 2.5 hours prior to entry interface, and CM power-up commenced on *Odyssey*'s batteries. In what was known as a reverse docked course alignment, the platform in *Odyssey* would be aligned with the platform in *Aquarius* using the LM thrusters to hold attitude. This was the reverse of the process used to transfer the *Odyssey* REFSMMAT to *Aquarius* shortly after the accident.

Mission Control could read the LM alignment through the IMU angles and verbally inform the crew so that they could place the numbers in the CMC for a course alignment. This would have been sufficient for re-entry purposes with a star check using Vega and Altair for accurate platform alignment. These numbers were passed to the crew and an IMU course alignment and a celestial check was made by 140:55 GET, obviating the need for a Sun and Moon alignment.

With fine alignment complete Apollo 13 was cleared to move to LM jettison attitude. Under normal circumstances the LM would be jettisoned around the Moon and the Service Module thrusters would have been used to back away from the Ascent Stage. But the Command Module was on its own, having separated from the SM but still attached to the LM, and had no independent means of translating away from *Aquarius*.

To ensure the two modules were well apart at re-entry, the LM would be shot from the top of the Command Module like a cork from a bottle. To produce that effect, after closing the hatch to the LM on the one side and the hatch at the top of the Command Module on the other, the docking tunnel would be kept pressurised at 2.2lb/in^2 (15.17kPa). Because the forward hatch on *Odyssey* was only 4in (10.2cm) below the pyrotechnic separation ring that would fire to disengage the two vehicles, there was concern that overpressure could crack or split the hatch.

The tunnel would normally be evacuated to a vacuum, but the technique had inadvertently been tried on Apollo 10. Grumman called world-renowned specialists at the University of Toronto, Institute of Aerospace Studies, to ask for detailed assessments of the forces involved. Using parametric data from the first atomic explosion at Alamogordo, New Mexico, they calculated the optimum pressure Δ and gave that to Grumman, who passed it on to NASA.

When the separation latches holding the docking ring were fired from inside the CM the effect would impart a separation rate of about 2ft/sec (0.61m/sec), a modest 1.4mph (2.25kph). In this way the two spacecraft would

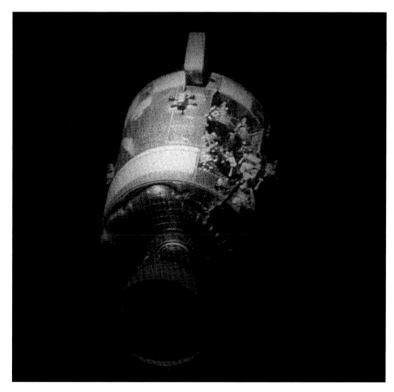

BELOW **An end view of the Service Module shows some discolouration on the SPS engine bell, with damage to the S-band antenna. Note the fly-away umbilical shroud hinged back after disconnecting from the Command Module.** *(NASA)*

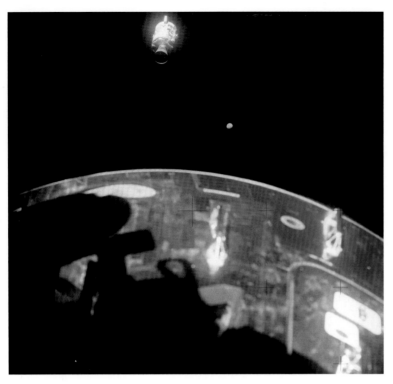

be at least 4,000ft (1,219m) apart at re-entry, but most probably double that.

However, when the crew steered the docked configuration to the jettison attitude, which was supposed to be at a roll of 228°, a pitch of 125° and a yaw of 12.4°, while correctly positioning the X-axis along the positive radius vector they put in a yaw of 45° south of the trajectory instead of 65° north, a coupled roll error of 110°. At 141:06 GET, RETRO 'Bobby' Spencer saw the telemetry indicating this and advised Gene Kranz that the alignment was incorrect. Before vacating *Aquarius* the crew left it in AGS-controlled attitude-hold with wide deadband errors to hold the LM/CM configuration stable up to separation.

By now the crew were closing up the tunnel to *Aquarius* and time was running out. Kranz decided that although the initial lift vector would carry the Command Module north on its re-entry ground track, it would be away from the LM orbit plane as it re-entered. The incorrect attitude would be no problem. The crew were ahead of the timeline as they ensured anything that did not have to come back was left in *Aquarius*, and the crew reassembled in *Odyssey* before closing the LM hatch for the last time.

At 141:05 GET Swigert called down to

Mission Control: 'OK, we're ready to proceed with hatch close up.' The forward docking hatch on *Odyssey* was shut and sealed. To check that it was pressure-tight, the docking tunnel between the two vehicles was pressurised to 2.8lb/in² (19.3kPa), with *Aquarius* at 4.98lb/in² (34.34kPa). The crew watched for several minutes to see if the tunnel pressure would rise, indicating a leak from the pressurised Command Module. There was none. Twelve minutes later Joe Kerwin gave a 'go' for LM jettison at the crew's discretion. Tension was high, biomedical harnesses now on Swigert and Haise showing heart rates in excess of 100 beats per minute.

At launch, nearly six days earlier, *Aquarius* had 18,434.5lb of propellant in the Descent Propulsion System tanks of which 18,338lb (8,318kg) was usable. A total of 8,342.9lb (3,784.3kg), or 45%, had been consumed in the three burns made by the LMDE. At separation the Ascent Stage RCS thruster systems had 166lb (75.3kg) of propellant remaining, 26% of the 633lb (287.13kg) loaded for launch. Of the 54lb (24.5kg) of oxygen loaded, 21.9lb (9.93kg) had been consumed in about 84 hours, all from the Descent Stage tank, with sufficient remaining (59%) from all three tanks for a further 135 hours.

Electrical usage had soared since the LM had been powered up and by additional loads placed on its circuits by the Command Module until it was independently powered by its re-entry batteries. When *Aquarius* ended its own unique mission there remained an estimated 410Ah – sufficient, at a demand rate of 38.8A, to last a further 11 hours. But the most immediate effect had been on water consumption.

Powering up the systems and putting power into *Odyssey* raised demand from 2.5lb/hr (1.13kg/hr), which except for the mid-course power-up it had maintained since 91:00 GET, to 5.4lb/hr (2.45kg/hr). At this elevated rate there was sufficient water for a further six hours at most. It had been a close-run thing. At the end of its useful life the LM weighed 24,730lb (11,217kg), about 74% of its launch weight.

The Command Module separated from *Aquarius* at 141:30:00.2 GET at a distance of 12,955 miles (20,848km) and a velocity now up to 11,908mph (19,160.4kph), just 70 minutes prior to entry interface with a gamma of -6.269°. As the Lunar Module drifted slowly away, majestic and dignified in its demise, Joe Kerwin spoke to the ether and expressed the sentiments of everyone at NASA as well as the three crew members of Apollo 13: 'Farewell *Aquarius*, and we thank you.' Seven minutes later Kerwin read up the final re-entry pad to Jim Lovell, defining the final tracking parameters for descent events.

Recovery

With its precious cargo of three very tired crew members, the Command Module was now on its own, irrevocably committed to slicing into the Earth's atmosphere. The RCS system was capable only of attitude orientation and had no capacity for altering the trajectory. That would come once the conical module felt aerodynamic pressure and could exercise a degree of lift vector to control the flight path. But everything depended now on the thrusters' ability to maintain attitude changes during descent until the parachutes took over.

Like the 16 thrusters attached to the four Service Module quads, the 12 thrusters in the Command Module operated on hypergolic MMH and N2O4 propellants. Manufactured by Rocketdyne, the CM thrusters were scarfed into the contour of the Command Module with ablative nozzle extensions, and delivered a thrust of 93lb (413.7N). Each thruster was 11.65in (29.6cm) long with an exit diameter of 2.12in (5.4cm) and weighed 8.3lb (3.76kg).

The CM RCS units were divided into two parallel systems for redundancy, designated System 1 and System 2, each with its independent fuel, oxidiser and helium pressurisation tank. The fuel and oxidiser tanks were identical to the equivalent tanks in the secondary system in the Service Module RCS system manufactured by Bell. As loaded on CSM-109, System 1 had 122lb (55.34kg) of propellants while System 2 had 123.1lb (55.84kg). Up to communication blackout during re-entry, 12lb (5.44kg) of propellant had been used by *Odyssey* – a large amount for a normal mission but not for the protracted time since separation from *Aquarius*, and the fact that the thrusters were used during that event in an unusual manner.

ABOVE As *Aquarius* is jettisoned, it slowly floats away, destined for destructive re-entry like the Service Module. *(NASA)*

Life support aboard *Odyssey* was controlled by the ECS. For the post-separation portion of the flight, the cabin was pressurised with pure oxygen from the surge tank in the CM. This was a 13in x 14in (33cm x 35cm) Inconel cylinder with a weight of 8.86lb (4kg) holding 3.77lb (1.71kg) of oxygen at a maximum pressure of 900lb/in² (6,205kPa). It was located in the left-hand equipment bay. The surge tank would normally provide pressurisation and for the crew to breathe for the period between separation of the Service Module and splashdown, usually about 30 minutes. In the case of Apollo 13 this would be closer to 90 minutes from the time *Aquarius* was cut loose.

In addition, three repressurisation bottles each containing 1lb (0.454kg) of oxygen were installed in the CM below the main hatch for use through the surge tank. They were capable of achieving a pressure of 3lb/in² (20.69kPa) from zero within two minutes. Also in the CM was an aluminium potable water tank with a diameter of 12.5in (31.75cm), weighing 7.9lb (3.58kg) and holding 4.25 gallons (16.1 litres) of water. At the time of the accident, the CM had 38lb (17.24kg) of water available, but 14lb (6.35kg) of that had been transferred to *Aquarius* for drinking and for food preparation.

ABOVE Technicians at Avco inject phenolic epoxy resin into a stainless-steel honeycomb matrix, to form an ablative protection for the Command Module. Although the base of the conical module experienced the most intense effects of re-entry, the ablative shield covered the entire exterior apart from the very apex where the forward hatch was located. *(Avco)*

BELOW The 370,000 separate honeycomb cells in 40 individual panels served as an effective thermal barrier. *(North American Rockwell)*

Manufactured by the Avco Corporation, the heat shield was an ablative compound of phenolic epoxy resin set hard in a stainless steel honeycomb matrix. It was required to survive temperature extremes of up to 1,200°F during ascent through the atmosphere, +/-280°F when facing the Sun or in shadow and up to 5,000°F during peak heating on re-entry. Behind the heat shield, insulation on top of a bonded aluminium honeycomb helped keep temperatures within comfortable limits on the pressurised interior of the module, usually 70°–75°F.

The thickness of the shield varied from 2.75in (7cm) at the base of the conical module to 0.25in (0.64cm) around the top of the apex. With a weight of 3,000lb (1,361kg) the shield accounted for almost one-third of the re-entry mass of *Odyssey*. A protective film of 0.5mil (0.0005in or 0.0127mm) Kapton coated with 0.00005in aluminium foil with a coating of silicon oxide came in the form of tape which was wound round the exterior of the Command Module. Much of it flaked away during re-entry leaving a charred and blistered appearance.

Because the module was designed to control its descent path during re-entry, flying almost parallel to the Earth for much of the time, soaring higher again into the upper atmosphere to reduce the heat rate before dipping down again to reduce speed, it had an active control system in the RCS thrusters. None of this could alter the heat load, which was determined by the mass and the entry velocity, but the heat rate could be kept within shield design limits.

Unlike the clipped edge of the Mercury and Gemini spacecraft, Apollo had a rounded rim to the convex lower heat shield. This was to prevent the shock wave attaching to the edge of the shield and pushing temperatures higher than necessary when the CM pitched up or down and moved the airflow closer to the edge. The shock wave would remain attached to the tip of the cone if the half-angle was less than 60°, and this angle (of attack) would be below that at some pitch levels, assuming the included angle of the base and the angle of the cone sidewall constituted a half-angle presented to the air.

There was also the question of heat shield integrity. Had it been damaged in the accident, clearly related to the sector containing fuel cells and cryogenic tanks? Had the explosion within the bay that ripped off the outer panel also directed a penetrating force thrusting up though the forward bulkhead and into the heat shield on the Command Module? The forward portion of the Service Module could not be seen as it drifted away, only the aft end and the side affected by the accident. Fred Haise remarked that the SPS engine bell appeared scarred but was effectively intact, while the S-band antenna had clearly been struck and damaged by the sidewall ripping loose.

The crew had jettisoned *Aquarius* about ten minutes earlier than planned, which gave them a little longer to prepare for entry. To enable some verification of IMU alignment to be made in the absence of an Earth horizon check, a Moonset check was planned. The celestial alignments of Earth and Moon placed the lunar disc in the centre of the No 2 (left docking) window directly in front of Jack Swigert. The window was scribed with markings and the crew were to control the attitude of the Command Module so that the Moon would hold at the 36° mark until Moonset at about EI-2.5 minutes, more precisely 142:38:19 GET. For re-entry Swigert occupied the left couch with Lovell in the middle and Haise in the right couch.

Entry interface would occur at 142:40:45.7 GET over a position 28.23°S by 173.44°E, south-east of New Caledonia, due east of Brisbane, Australia. From the point of entry at a height of 400,000ft (121,920m), or 75.76 miles, the Command Module would fly a lifting trajectory with maximum deceleration of 5.2g for a total distance of 1,288.6 miles (2,073.3km) to splashdown. It would slow from 36,291ft/sec (11,061m/sec), or 24,743mph (39,812kph) to a mere 31ft/sec (9.45m/sec), or 21mph (34kph), as it sliced into the Pacific Ocean almost 14 minutes later.

During the early part of the descent, when the Command Module was experiencing maximum heat rate, a plasma sheath would build up which would block radio waves, a period known as the communications blackout. The final updates to the EMS showed beginning of blackout 19sec after EI and end of blackout

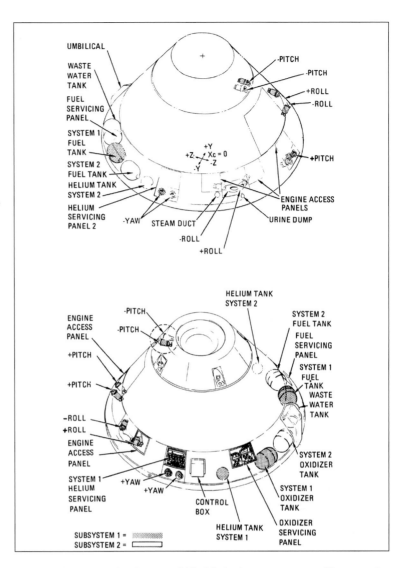

ABOVE Alone now, the Command Module had an autonomous life-support system and an attitude-control system for orientation during the approach to re-entry and during the passage through the atmosphere where it would control the lift vector for accurate splashdown. These diagrams show the RCS system and associated propellant tanks and delivery systems. The propellant tanks were identical to the secondary tanks in the Service Module RCS units, while the four titanium helium tanks had a diameter of 9.2in (23.37cm) with a wall thickness of 0.102in (0.26cm) and a capacity of 0.57lb (1.45kg), with a pressurisation at 4,150lb/in² (28,614kPa). *(North American Rockwell)*

at 3min 38sec. The drogue parachutes were predicted to deploy at 7min 59sec followed by the main parachutes at 8min 46sec, with splashdown 4min 58sec later.

In the final minutes before splashdown the mood in Mission Control was electric with tense, expectant emotion, pent up and controlled but palpable in its energy. Just about every element of the NASA hierarchy was in the MOCR or behind the big glass screen in the viewing room. It was as though the entire future of the space programme depended on these three men finally making it back to Earth. And in some ways it did. But while the management teams were drifting about, chatting among themselves or conferring intently, the White Team was calm, cool and collected.

Gene Kranz was pensive and intently fixed on his screens, looking across to various consoles as the trenches did their work and watched – and waited. Eyes never blinking, fixed and wide, an expressionless visage, Kranz knew that this was the deciding moment that would determine whether NASA survived its greatest failure or dominated world headlines with news of a disaster.

With 18 minutes remaining before re-entry, Jack Swigert called Joe Kerwin with a gesture of gratitude: 'I know all of us here want to thank all you guys down there for the very fine job you did.'

Lovell added an endorsement: 'That's affirm, Joe.'

And from Mission Control: 'I'll tell you – we all had a good time doing it.'

With 11 minutes to re-entry, a little levity from bachelor Jack Swigert, anticipating the eruption that would overwhelm Mission Control when the crew got down safe: 'Sure wish I could go to the FIDO party tonight.'

Kerwin, with a background of laughter on the airwave, replied: 'Yes, it's going to be a wild one...somebody said "We will cover for you guys, and if Jack's got any phone numbers he wants us to call, why, pass them down!"'

Gene Kranz lit another cigarette and, standing up to survey his controllers, went around the room calling each in turn for a 'go' or 'no go', which was a formal way of signing off to Isaac Newton and the law of gravity. There were no dissenters. Nothing more could be done. Aboard *Odyssey*, the mass of planet Earth loomed large and dominant out the windows, as, heads down, their backs to the direction of motion, they flew across the curving horizon into darkness, checking perfectly the alignment of the Moon through the left docking window.

The last word from Joe Kerwin came just prior to re-entry: 'OK, LOS [loss of signal] in a minute or a minute and a half, in entry attitude we'd like Omni [antenna] Charlie – and welcome home, over.' As they entered the atmosphere a faint glow began to surround the Command Module, turning brilliant orange as the fireball outside enveloped the tiny capsule, the heat shield giving up its ablative material to the searing heat, eroding as planned until the worst was over and the spacecraft had slowed sufficiently for the glow to subside.

Had all gone well the communications blackout should have ended 3min 38sec after entry interface, but when the call went up from Joe Kerwin at Mission Control there was only a hissing sound coming through the airwaves. It was only now that the world would learn whether *Odyssey*'s heat shield had remained intact and undamaged, and it was not unusual for a lapse in re-establishing contact after blackout. But of all missions, on this one the silence was deafening.

Again the call went up from Joe Kerwin, now 30 seconds after end of blackout: '*Odyssey*, Houston standing by, over.' Still there was nothing but silence. Seconds more went by: '*Odyssey*, Houston standing by, over.' Nothing but silence. A minute had now gone by.

ONE-PIECE TRIMMING ORIFICE
FILTER
FUEL VALVE
OXIDIZER VALVE (ROTATED 90° & SECTIONED)
ABLATIVE SLEEVE
STAINLESS STEEL SHELL
ABLATIVE MATERIAL
ASBESTOS WRAP
CUSHION
REFRACTORY THROAT INSERT
GLASS WRAP

The call went out a third time and Gene Kranz stood up, stubbing out his cigarette, taking worried glances around the MOCR. Then suddenly, with the signal monitor flickering, a voice from Apollo 13, 1min 39sec after end of blackout: 'OK, Joe,' and Jack Swigert signalled that the crew had indeed survived re-entry. Kranz beamed broadly and punched the air with his fist, only those close at his side detecting moist eyes in an otherwise iron expression of restrained joy and delight.

And from the Capcom console: 'OK, we read you, Jack,' after which Swigert confirmed that the two drogue parachutes had deployed. Falling at a terminal velocity of 300mph (483kph), at 24,000ft (7,315m) the forward heat shield, or apex cover, was jettisoned, exposing the parachute recovery section and falling away toward the sea. Two seconds later, at about 23,000ft (7,010m), two 16.5ft (5m) white nylon conical ribbon drogue parachutes were deployed, ejected by mortars at the top of the module triggered by release of the apex cover. Extended 65ft (19.8m) above the Command Module, they slowed the fall rate to 175mph (281kph), followed 47sec later by deployment of three white nylon ring-slot pilot parachutes. With a diameter of 7.2ft (2.2m), these reached a height of 58ft (17.7m) above *Odyssey*, each pulling out a single 83.5ft ringsail main parachute. Deployed at an altitude of 10,000ft (3,048m), they were suspended 120ft (36.6m) above the Command Module, slowing the descending craft to 21mph (33.8kph).

For almost five minutes *Odyssey* fell toward the sea, drifting only slightly in the calm Pacific air, half a world away from the place where they had departed the Earth on the most perilous space voyage ever taken. As it descended the module was pitched 27.5°, so that it would slice into the sea at an angle rather than slap down in a bone-jarring belly-flop. Below them the sea was calm, with a gentle swell, light winds, good visibility and only a few scattered clouds.

In Mission Control the picture on the big TV screen to the right of the plot-board across the front wall of the MOCR now showed a tiny conical spacecraft descending slowly through blue Pacific skies. From Joe Kerwin: '*Odyssey*, Houston. We show you on the mains. It really looks great...Got you on television, babe.'

ABOVE The main components of the Earth-landing system include two drogue mortars, seen here as two cylinders facing outward just above the two pitch thrusters. The main parachutes are packed around the side and back in this view. *(North American Rockwell)*

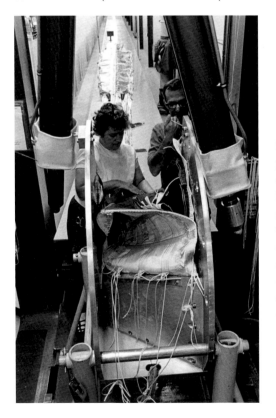

LEFT All the parachutes on *Odyssey*'s Command Module were manufactured by the then Northrop-Ventura company, and were packed using a special compression ram to encapsulate the broad expanse of folded canopy into the limited space around the apex of the Command Module. *(Northrop-Ventura)*

ABOVE Apollo 13 appears in the high clouds above the Pacific following a suspenseful re-entry and communications blackout that left nail-biting tension in Mission Control. *(US Navy)*

RIGHT The Command Module had crushable ribs acting as protection for toxic propellant tanks that would move with slight deformations in the structure when it struck the water. *(North American Rockwell)*

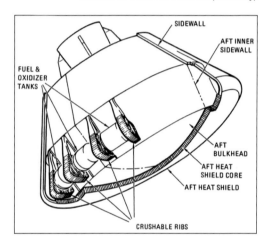

Across Mission Control – thick with more than 100 people, exultant, thankful, some prayerful, many with smiles across grained faces etched with fatigue, relief, exhaustion and now exuberance – there came an outpouring of emotional release, as people were suddenly, almost in disbelief, released from a burden that had descended upon them more than three and a half days ago. Days in which few had been able to do more than snatch catnaps to relieve the pressure.

Splashdown occurred at 142:54:41 GET, a little more than one hour short of six days after launch from the Kennedy Space Center. In Houston it was approaching 12:08pm on Thursday 17 April, but in the Pacific Ocean west of the Cook Islands it was 7:08am. The international response to Apollo 13 had been unprecedented, with 13 ships and 17 aircraft from several countries pledged to support a Pacific landing, if needed. In addition, while a day after the accident there was still uncertainty as to where *Odyssey* would come down, 21 ships and 17 aircraft had been pledged for the Indian Ocean and 51 ships and 21 aircraft for a landing in the Atlantic Ocean. But none of these were needed.

BELOW Splashdown! For many it was an event that raised greater emotional release than did the first landing on the Moon. NASA had saved three family members, and nothing could beat the exuberance of that. *(NASA)*

BELOW The Command Module had an uprighting system with three 'golf-ball' bags that could be inflated manually by switching on two air compressors in the aft compartment. *(North American Rockwell)*

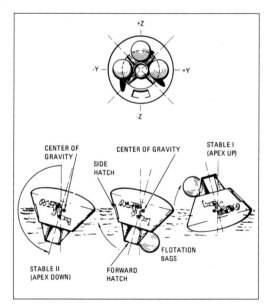

At the time of landing, the prime recovery vessel USS *Iwo Jima* was sitting just 5.75 miles (9.25km) north-west of the touchdown spot at 21° 34.7'S by 165° 23.2'W. With the descent path trending south-west to north-east, *Iwo Jima* was to the left of the ground track. Designated Task Force 130, the recovery force comprised the *Iwo Jima*, five of its helicopters and four Lockheed HC-130H rescue aircraft.

At splashdown the bridge officer conferred with the navigation officer, who calculated a range of 4.55 miles (7.32km), with the Command Module on a bearing of 158.9° east of north. *Odyssey* was sitting in the water at 21° 38' 24"S latitude by 165° 21' 42"W longitude, about 3,000ft (914m) from the calculated touchdown point.

One helicopter, a Sikorsky SH-34 designated Recovery 1, carried a flight surgeon and was assigned to retrieve the crew and get them on the deck of the *Iwo Jima*. Piloted by Commander 'Chuck' Smiley, who had recovered the crew of Apollo 10 eleven months earlier, it carried a co-pilot, two crew members and a swimmer. Waiting over the *Iwo Jima* until splashdown, Recovery 1 would hoist the crew aboard using a Billy Pugh net, used for all previous recoveries but modified for this mission by aluminium stiffeners to make it more comfortable.

Two other helicopters were designated Swim 1 and Swim 2, the former piloted by Lieutenant Commander Carl Frank with a co-pilot, two crew members and three swimmers. Swim 2 was piloted by Lieutenant Allen Willhite and had a similar complement. The swimmers were to jump from the helicopter and deploy a flotation collar around the Command Module, inflate a dinghy and prepare for retrieval by Recovery 1. Swim 1 had been waiting 11.5 miles (18.5km) up-range and 17.3 miles (27.8km) north of the *Iwo Jima*, with Swim 2 at equal distances down-range of ship.

A fourth helicopter, designated Photo, carried camera crew, while the fifth, Relay, would provide a communication bridge between the airborne forces and the ship. Both helicopters hovered over the recovery ship until splashdown, and it was from Photo that most of the post-splashdown images came.

Launched on 17 September 1960, *Iwo Jima* was an amphibious assault ship commissioned on 26 August 1961, and had begun the first of six deployments to the western Pacific in September 1963. In April 1970 its skipper,

LEFT The recovery ship USS *Iwo Jima*, never a better sight for three men who would now be remembered by history as achieving fame to equal the first Moonwalkers.
(US Navy)

RIGHT A US Navy
SH-53 snatches the
crew from the Pacific
Ocean in modified Billy
Pugh nets. *(US Navy)*

who would welcome Apollo 13's crew aboard, was Captain Leland E. Kirkemo. In addition the USS *Granville Hall* had been standing off west-south-west in the event that a 'constant-g' re-entry had to be flown, foreshortening the down-range distance.

The HC-130H aircraft were designated Somoa Rescue 1, 2, 3 and 4, and it was Rescue 4 that had first reported an S-band contact with *Odyssey* six minutes prior to splashdown. Visual contact was acquired by Swim 2 a minute later, with voice contact through Recovery 1 a minute after that. The helicopters arrived over *Odyssey* at 7:16am local time, about eight minutes after splashdown.

About two minutes later the uprighting system was activated, a set of three 22ft³ (0.62 m³) inflatable air bags, designed to right the module should it roll over to an apex-down position. The conical CM is stable in No 1 (upright), or No 2 (inverted) position, and the air bags would right the module to the No 1

position if waves or parachute lines should tip it. *Odyssey* remained in stable mode 1.

Swim 2 deployed its swimmers to attach a flotation collar around the conical module while Swim 1 went off to retrieve the apex cover and to do what it could to hook up the three main parachutes. (The apex cover is jettisoned during descent and exposes the parachute containers.)

The hatch to *Odyssey* was opened at 7:32am and all three crew members were in the dinghy three minutes later, to be winched aboard Swim 2 at 7:42am, Fred Haise being followed by Jack Swigert and finally Jim Lovell. At 7:52am they were standing on the deck of the *Iwo Jima* being welcomed aboard by Captain Kirkemo, who said to the three men, 'I want to commend you on your navigation. Welcome aboard the *Iwo Jima*.'

A short prayer of thanksgiving was given by the chaplain, then they were quickly on to the side elevator for a few photographs and down to the hangar deck, where they made their way a

short distance across to the medical facility. Three doctors were assigned to a brief examination of each astronaut. An estimate based on the medical debriefing indicated that all three crew members were heavily deprived of sleep.

When the accident happened at 55:55 GET they had been awake for more than nine hours and were scheduled to start their next sleep at 61:00 GET. None had managed any rest until around 80:00 GET. In the span of 96 hours between their pre-accident rest and splashdown, Lovell managed a total of 11 hours, Swigert 12 hours and Haise 19 hours. None had consumed more than 24fl oz (0.71 litres) of water in the 87 hours since the accident.

As regards weight, Lovell had lost 14lb (6.35kg), Swigert 11lb (4.99kg) and Haise 6.5lb (2.95kg). In the final hours of the mission Haise

had consumed a great deal of water, more than the other crew members, but was confirmed to have a urinary tract infection. All three were diagnosed with extreme fatigue. Each had taken a Dexedrine tablet – a drug known as a stimulant and for improving focused attention – in the two hours prior to re-entry.

Meanwhile, the Command Module was retrieved at 8:36am and brought aboard on the main side elevator of the *Iwo Jima*. The apex cover and one parachute were retrieved by a small boat. At re-entry what remained of *Odyssey* had weighed 12,361lb (5,607kg), but on the water, without parachutes and some of its heat shield which had charred away as designed, it weighed 11,133lb (5,050kg) – all that was left of a combined mass of 110,252lb (50,010.3kg) at lift-off, including the boost

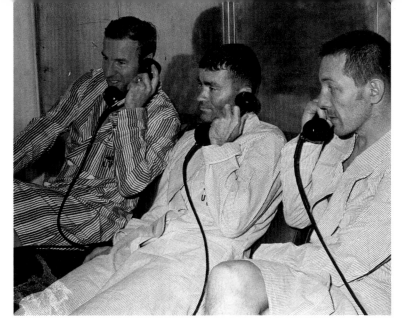

protective cover and the launch escape system, jettisoned during ascent.

The crew departed the *Iwo Jima* for Pago Pago, Samoa, by helicopter at 7:20am the following day, Friday 18 April. From there they were flown to Hawaii in a Lockheed C-141 transport aircraft, arriving at 4:35pm. They were greeted by President Nixon, who had brought along Marilyn Lovell and Mary Haise in the Presidential retinue. A ceremony was held followed by an overnight stay, after which they were flown back to Houston on 19 April, arriving at 10:30pm.

At 8:30am on 24 April the *Iwo Jima* docked at Hawaii, carrying the precious Command Module, which was taken to Hickam Air Force Base, where the pyrotechnics were deactivated and the fuel and oxidiser tanks made safe. It was then shipped in a Douglas C-133 to North American Rockwell at Downey, California, where it arrived at 9:00am on 27 April for an exhaustive inspection. But other parts of Apollo 13 lay at the bottom of the Pacific Ocean, along with pieces of the wrecked Service Module but little of *Aquarius* except for a graphite cask, 16.5in (41.9cm) long by 2.5in (6.35cm) wide, weighing 15.5lb (7kg) and containing 8.6lb (3.9kg) of plutonium 238 dioxide that would have powered the array of science instruments set out on the Moon. That lies north-east of New Zealand, 20,000ft (6,096m) down in the very deep Tonga Trench. No radiation has ever been detected.

ABOVE LEFT
Telephone calls to wives, loved ones and the President of the United States, Richard Nixon, occupy their first few hours. *(NASA)*

ABOVE **Jim Lovell catches up on the news – about himself and his crew. Afterwards he said how surprised he was to see the worldwide interest and support for their mission.** *(NASA)*

LEFT Richard Nixon awards the three astronauts the Presidential Medal of Freedom in the category of Space Exploration. *(NASA)*

Chapter Eight

What went wrong?

Had the accident to Apollo 13 happened when Apollo 8 went to the Moon in Christmas 1968, when no Lunar Module was available as a lifeboat, the crew could not have survived. So what caused this near disaster? And what could be done to prevent it happening again? Was it a random failure, or human error? NASA was eager to find out and get back to the Moon.

OPPOSITE Launched on 31 January 1971, Apollo 14 sailed the voyage aborted during Apollo 13, carrying astronauts Alan Shepard (centre), Stuart Roosa (left) and Edgar Mitchell to the rolling hills of the Fra Mauro site. *(NASA)*

The day Apollo 13 returned to Earth NASA administrator Thomas O. Paine and his deputy George M. Low set up a review board headed by Dr Edgar M. Cortright, director of NASA's Langley Research Center, to discover what caused the accident. When the final report was submitted on 15 June 1970 it listed a series of faults in the design and failures in procedures that led directly to what it described as a 'nearly catastrophic' event.

It found that an electrical fire in oxygen tank No 2 caused over-pressurisation and rupture which also damaged oxygen tank No 1, resulting in it leaking its contents into space. It determined that tank No 2 had not exploded, as evidenced by the pictures taken after Service Module separation, showing a major part of the tank intact, but that it had split as pressure exceeded its burst value.

There was no explosion, however dramatic that may have been and however many times it has been said subsequently in books and films.

The Board found that the design of the fans and heaters was susceptible to damage and noted the erratic and problematic history of the tank prior to final installation in the Service Module, as explained earlier in this book. It also found that there were combustible materials in the tank itself, such as Teflon insulation, but asserted that the quantity measurement probe did not have sufficient electrical current to cause a fire.

One damning piece of evidence was that changes made by NASA in electrical loadings imposed upon the tank in ground tests at the Kennedy Space Center were not acted upon. When Beech Aircraft was given the contract in 1962 to build the tanks they

were required to accept a loading of 28V DC. In 1965 Apollo prime contractor North American Aviation issued a requirement that the switches and power circuits associated with the tank must henceforth be capable of 65V DC, to accommodate changes made to ground test equipment at KSC.

This was not acted upon, or noticed, by Beech, NAA or NASA, and tanks continued to be delivered with internal switches and heaters that could overheat, partially melt or fuse under these increased loadings. All Apollo cryogenic tanks were thus vulnerable to a fire that would rapidly convert super-cold liquid oxygen into a gas, causing rupture.

The Board found this fact – together with bad handling during installation, combustible materials inside the sealed tanks and continued recycling of the fans and heaters during the early part of the mission – as the cause of the electrical short that resulted in the fire. Tests demonstrated that heater switches could have reached 1,000°F (538°C). The Board found that a short circuit in the fans opened wiring and caused one fan to fail completely, creating a short circuit of 10 joules, more than enough to ignite Teflon wire insulation.

It found through tests that the ruptured tank had quickly filled the sector 4 compartment with oxygen beyond the 25lb/in^2 (172.4kPa) limit of the external panel, causing it to blow off. This overpressure also played a significant role in damaging tank No 1, which also lost pressure and eventually leaked to depletion. Most frightening of all, however, was that a pressure of only 10lb/in^2 (68.9kPa) imposed upon the underside of the aft Command Module heat shield would rip the Service Module away completely.

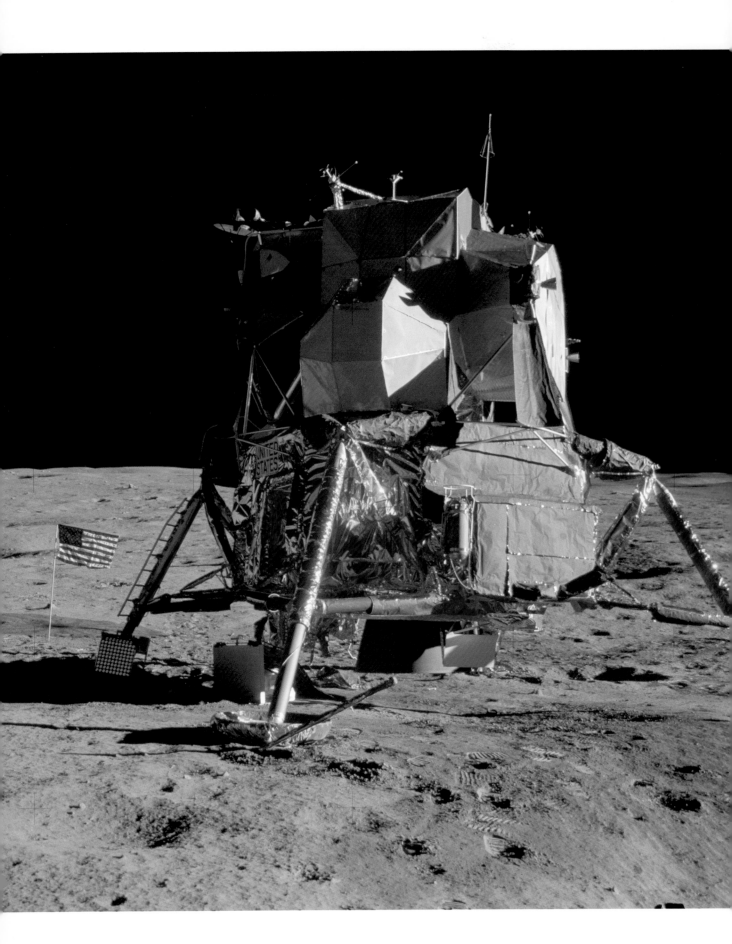

In the unlikely event that the gas pressure had ruptured the forward bulkhead it could have left the CM exposed to the extremes of temperature variations in space, and debate about whether to use *Aquarius* would have been moot. The loss of oxygen and electrical power, still available to the crew in the dying SM for more than two hours, would have been instant, leaving inadequate time to get aboard the LM and power it up.

Fixes

NASA was quick to make changes to tank design and enclosed all internal wiring in stainless steel conduits instead of Teflon insulation, adding a third heater element with optional switching of one, two or three elements instead of two on or off as previously. A sensor was added for the crew and ground control to access heater assembly temperatures, and the bulk temperature sensor was relocated. The quantity probe was changed from aluminium to stainless steel and soldered joints were replaced with brazed joints.

Changes also included the addition of a third oxygen tank, identical to the other two of the newly modified design. It was placed at the top of sector 1 on the opposite side of the Service Module to the cryogenic tanks for the electrical production system. Previously this sector had been vacant, but the lower section would carry a set of science instruments for Apollos 15–17.

To increase the available electrical supply, an auxiliary battery identical to the four LM Descent Stage batteries was installed in sector 4 of the Service Module, a 400Ah silver-oxide/zinc non-rechargeable type which had kept power going to Apollo 13 for so long.

Water capacity was increased by the addition of five plastic bags, each with a capacity of one US gallon (3.8-litre), wrapped in beta cloth and packed in a bag with fill hose, valves and a drinking nozzle, installed in a locker on the aft bulkhead in the Command Module. Tests showed that under some power-down circumstances the water in the storage tank could freeze, so the bags would be filled before that happened.

Consequences

Apollo 14 had been scheduled for October 1970 but that was put back to 3 December after the accident. But NASA's budget was being cut dramatically, and for reasons quite outside the Apollo 13 effect. On 2 September 1970 it cancelled two Moon missions, reducing the remaining total to four, and deferring Apollo 14 to 31 January 1971, the first mission to carry the post-accident modifications.

Flown in 1971 and 1972, Apollos 15–17 would use greatly modified spacecraft that had been in the planning stage for several years and would carry a newly commissioned Lunar Roving Vehicle, each visit lasting three days on the surface, with three separate periods of exploration each lasting 6–7 hours.

While giving NASA unbridled publicity at a time of public apathy and media disinterest, the drama of Apollo 13 was the last thing NASA wanted in 1970. It was retiring the Apollo programme already and trying to convince Congress to support a reusable launch system that would be known as the Shuttle, and to build a permanent habitation in space, which would eventually become the International Space Station. It did not want to expose the inherent flaws, imperfect design and lack of quality inspection implied by the Review Board.

Apollo 13 came only 38 months after the devastating fire at Cape Canaveral that had taken the lives of astronauts Virgil 'Gus' Grissom, Edward White and Roger Chaffee. While acknowledging that returning the Apollo 13 crew safely to Earth was a human triumph it was further evidence that human space flight would always be dangerous, and a reminder that NASA's glory days were over, that the aura that surrounded the first landing on the Moon – less than eight months before Apollo 13 – had become tenuous.

Apollo 13 revealed that difficult missions are balanced on a knife-edge between success and catastrophe, and that the future would be hard, challenging and contested as much by those who questioned America's right to journey forth on the new frontier, demolishing the budget as they went, as by the hard taskmasters of science and engineering.

Appendix A

The Command and Service Modules

The Apollo spacecraft assigned to Apollo 13 (CSM-109) consisted of two modules, a pressurised Command Module (CM) and an unpressurised Service Module (SM), held together until separated just before re-entry at the end of a mission. The SM housed electrical power production equipment, propulsion for attitude control and orbit changes, oxygen for the CM and communications equipment.

The SM was a barrel-shaped structure supporting the CM at the top and was characterised by the engine skirt for the Service Propulsion System (SPS) at the bottom. It had a diameter of 12ft 10in (3.91m) and a length of 12ft 11in (3.94m). A circular

RIGHT The Apollo 9 spacecraft depicts the assembled configuration of Command Module and Service Module, the former wrapped in blue tape insulation as a protection for the heat shield. (North American Rockwell)

extension of the structure comprised a fairing 22in (55.9cm) in height to enclose the space between the top of the SM and the CM. Including this and the SPS skirt the SM was 24ft 2in (7.34m) tall.

The SM was built up from forward and aft bulkheads with the interior divided into six sectors, with a central tunnel 44in (111.8cm) in diameter. The radial beams dividing the six pie-shaped sectors were fabricated from solid aluminium alloy, chemically milled to varying thicknesses between 2in (5cm) and 0.018in (0.05cm) for weight reduction. Forward and aft bulkheads provided a sealed structure, with projection of the forward beam trusses to provide fixtures for attaching to the Command Module at three locations, where tension ties held the two modules together.

The six sectors were of three separate widths: sectors 1 and 4 of 50° radius, sectors 3 and 6 of 60° radius and sectors 2 and 5 of 70° radius. Sector 1 was empty, originally assigned to carry cameras when Apollo was expected to conduct Earth studies, or a telescope for studying the Sun. Sector 2 enclosed the sump tank for nitrogen tetroxide oxidiser for the SPS engine and Sector 3 contained the oxidiser storage tank. Sector 5 contained the sump tank for unsymmetrical dimethylhydrazine fuel for the SPS engine while sector 6 contained the storage fuel tank.

Sectors 2, 3, 5 and 6 each supported a Reaction Control System (RCS) thruster quad on the outer surface, located 90° apart. Each sump tank was 153.8in (390.6cm) tall and 51in (129.5cm) in diameter. Each storage tank was 154.5in (392.3cm) high with a diameter of 45in (114.3cm).

About 75% of the weight of the SM was made up of propellants for the SPS engine

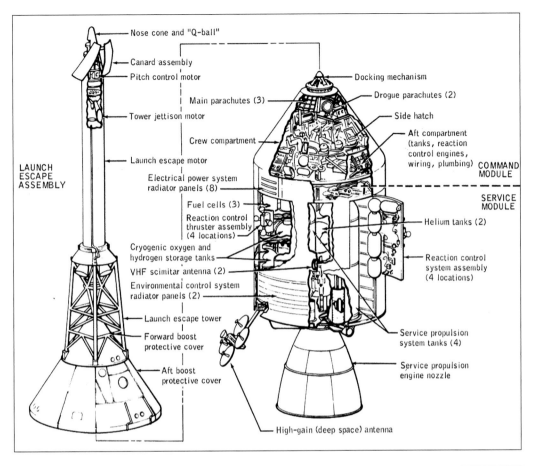

Nose cone and "Q-ball"

Canard assembly

Pitch control motor

Tower jettison motor

LAUNCH ESCAPE ASSEMBLY

Launch escape motor

Electrical power system radiator panels (8)

Fuel cells (3)

Reaction control thruster assembly (4 locations)

Cryogenic oxygen and hydrogen storage tanks

VHF scimitar antenna (2)

Environmental control system radiator panels (2)

Launch escape tower

Forward boost protective cover

Aft boost protective cover

Main parachutes (3)

Crew compartment

Docking mechanism

Drogue parachutes (2)

Side hatch

Aft compartment (tanks, reaction control engines, wiring, plumbing)

COMMAND MODULE

SERVICE MODULE

Helium tanks (2)

Reaction control system assembly (4 locations)

Service propulsion system tanks (4)

Service propulsion engine nozzle

High-gain (deep space) antenna

LEFT The docked Command and Service Modules are stacked on top of the Saturn V for launch. The Launch Escape System (LES) is carried through second stage ignition, should it be necessary to lift the Command Module away from a booster going out of control or in imminent danger of blowing up. The CM windows and exterior heat shield are protected from the effects of the atmosphere during ascent, and from the exhaust products of the LES jettison motor, by a boost protective cover that comes away with the jettisoned LES. *(North American Rockwell)*

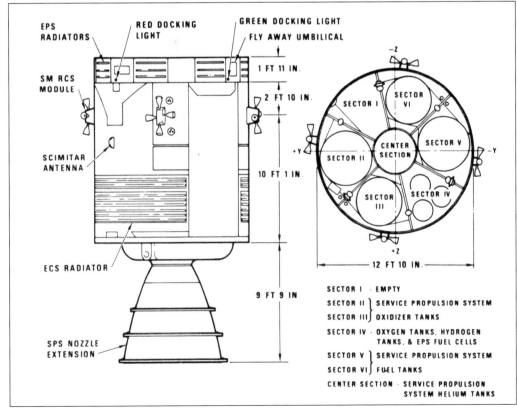

EPS RADIATORS

RED DOCKING LIGHT

GREEN DOCKING LIGHT

FLY AWAY UMBILICAL

SM RCS MODULE

SCIMITAR ANTENNA

ECS RADIATOR

SPS NOZZLE EXTENSION

1 FT 11 IN.

2 FT 10 IN.

10 FT 1 IN.

9 FT 9 IN

12 FT 10 IN.

SECTOR I
SECTOR VI
SECTOR V
CENTER SECTION
SECTOR II
SECTOR III
SECTOR IV

-Z
+Y
-Y
+Z

SECTOR I - EMPTY
SECTOR II } SERVICE PROPULSION SYSTEM
SECTOR III} OXIDIZER TANKS
SECTOR IV - OXYGEN TANKS, HYDROGEN TANKS, & EPS FUEL CELLS
SECTOR V } SERVICE PROPULSION SYSTEM
SECTOR VI } FUEL TANKS
CENTER SECTION - SERVICE PROPULSION SYSTEM HELIUM TANKS

LEFT Design layout of the Service Module shows the six sectors, four of which are for propellant tanks for the Service Propulsion System, one of which carries the fuel cells and cryogenic oxygen and hydrogen tanks, and one which is empty. The empty sector 1 space was utilised on the Apollo 15–17 missions for science instruments. *(North American Rockwell)*

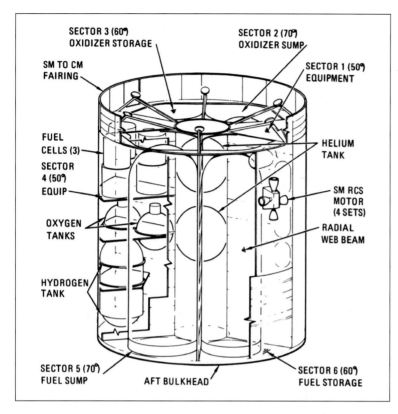

SECTOR 3 (60°) OXIDIZER STORAGE
SECTOR 2 (70°) OXIDIZER SUMP
SM TO CM FAIRING
SECTOR 1 (50°) EQUIPMENT
FUEL CELLS (3)
HELIUM TANK
SECTOR 4 (50°) EQUIP
SM RCS MOTOR (4 SETS)
OXYGEN TANKS
RADIAL WEB BEAM
HYDROGEN TANK
SECTOR 5 (70°) FUEL SUMP
AFT BULKHEAD
SECTOR 6 (60°) FUEL STORAGE

LEFT The Service Module comprised upper and lower bulkheads, six radial beams, four sector honeycomb panels, four RCS honeycomb panels, an aft thermal-protection panel, and a fairing around the top. Radial beams were fabricated from solid aluminium alloy machined and chemically milled to between 2in (5.08cm) and 0.018in (0.05cm). The aft thermal protection shield enclosed the SM from heat generated by the Service Propulsion System during firing. The fairing is 22in (55.9cm) high, and comprised 16 sections, eight of which were honeycomb panels alternating with eight radiators for the electrical power system. *(North American Rockwell)*

BELOW The SM was adaptable due to its segmented design, and had great potential for modification and improvement, as evidenced by the changes made as a result of the Apollo 13 incident. *(North American Rockwell)*

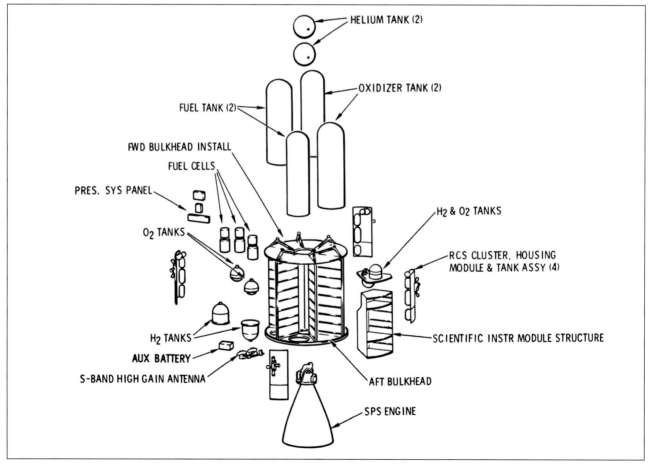

HELIUM TANK (2)
OXIDIZER TANK (2)
FUEL TANK (2)
FWD BULKHEAD INSTALL
FUEL CELLS
H2 & O2 TANKS
PRES. SYS PANEL
O2 TANKS
RCS CLUSTER, HOUSING MODULE & TANK ASSY (4)
H2 TANKS
SCIENTIFIC INSTR MODULE STRUCTURE
AUX BATTERY
S-BAND HIGH GAIN ANTENNA
AFT BULKHEAD
SPS ENGINE

RIGHT An Apollo Command Module in a fit check
with the top of the Service Module to which it
would be attached by three tension ties, and on
which it would be supported on six compression
pads. *(North American Rockwell)*

and the 16 RCS thrusters, but quantities varied
slightly with each mission. At launch SM-109
carried 15,685lb (7,115kg) of fuel and 25,084lb
(11,378kg) of oxidiser, for a total load of
40,769lb (18,493kg).

The four RCS quads comprised four
radiation-cooled, non-ablative thrusters with a
length of 13.4in (34cm) and an exit nozzle 5.6in
(14.2cm) in diameter. With solid molybdenum
thrust chambers and cobalt-alloy nozzles,
each motor delivered a thrust of about 100lb
(444.8N) for a duration anywhere between 12
milliseconds and 500 seconds, for a total cycle
of up to 10,000 firings. Propellants were the
same as those used in the SPS engine.

Each RCS quad had a primary and a

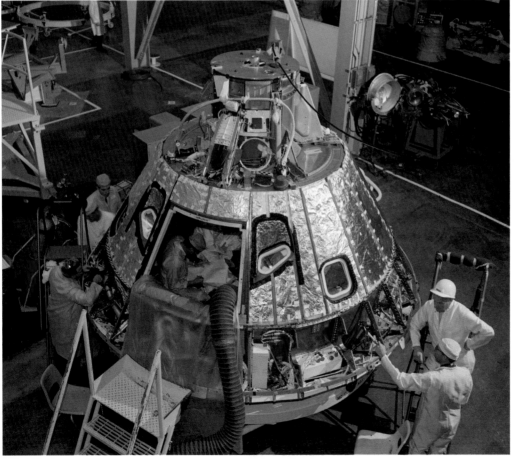

LEFT The inner
pressure shell of the
Command Module
consisted of an
aluminium-sandwich
construction with a
welded outer skin,
adhesively bonded
honeycomb core,
and face sheets.
The thickness of the
honeycomb varied
from 1.5in (3.8cm)
at the base to about
0.25in (0.6cm) at the
forward section near
the docking tunnel.
*(North American
Rockwell)*

ABOVE The primary structure was formed from aft, crew and forward shells, with the heat shield base (right) enclosing the aft section of the Command Module. *(Avco)*

ABOVE RIGHT Engineers at Avco Corporation fabricate the external heat shield, which when attached to the Command Module forms a shell over the pressure vessel. *(Avco)*

RIGHT Some equipment serving the Command Module's many systems was located between the inner pressure vessel and the heat shield, making it easier to service hazardous materials or toxic propellants for the RCS thrusters. *(North American Rockwell)*

secondary RCS propellant tank for fuel and oxidiser, each cylindrical with hemispherical end domes. These RCS quads operated on the positive expulsion technique, in which propellant was carried inside a Teflon bladder within the tank, the space between the bladder and the tank wall being pressurised with helium. Each spherical helium tank had a diameter of 12.4in (31.5cm) and contained 1.35lb (0.61kg) of helium.

The primary RCS oxidiser tanks were 25.6 x 12.6in (65 x 32cm) and the secondary tanks were 19.9 x 12.6in (50.6 x 32.1cm). The primary RCS fuel tanks were 23.7 x 12.6in (60.2 x 32cm), while the secondary tanks were 17.3 x 12.6in (43.9 x 32cm). SM-109 was loaded with 440.1lb (199.6kg) of fuel and 902.7lb (409.5kg) of oxidiser.

The Command Module was capable of supporting a crew of three for up to 14 days in space. It had the shape of a convex-bottomed cone, 10ft 7in (3.2m) tall to the tip of the docking probe with a diameter of 12ft 10in (3.91m). The conical sidewalls provided inset windows, affording ample view forward for docking with another vehicle at the apex end of the spacecraft. The top of the conical structure incorporated docking equipment and parachutes for recovery after re-entry through the atmosphere.

The CM comprised an inner, pressurised structure with a pressurised volume of 270ft³ (7.65m³) and a habitable volume of 210ft³ (5.9m³), and an outer structural shell, or heat shield, which comprised the full exterior of the module. The welded aluminium inner skin supported an adhesively bonded aluminium honeycomb core and outer face sheet, the thickness of which varied from 1.5in (3.8cm)

BELOW The configuration of the Command Module, showing the arrangement of various systems and elements. *(NASA)*

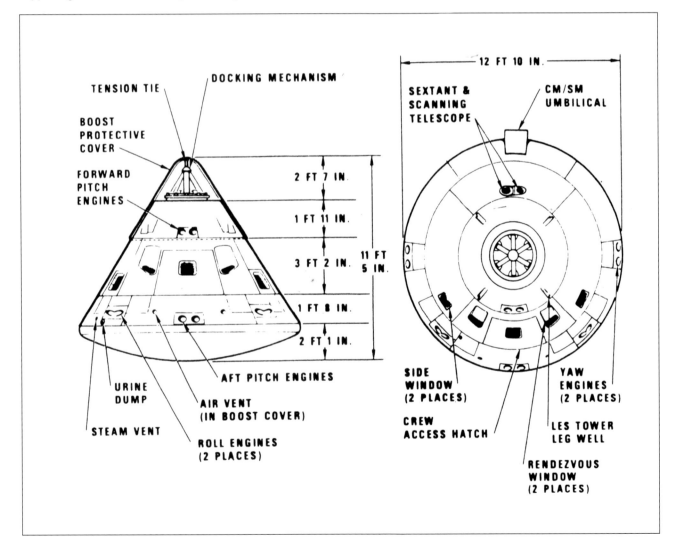

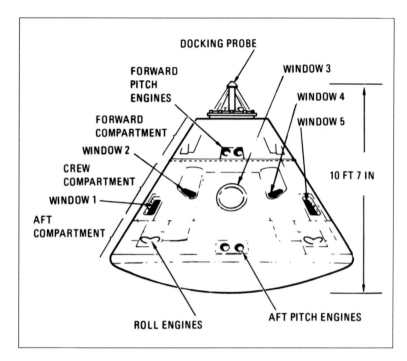

DOCKING PROBE

FORWARD PITCH ENGINES

FORWARD COMPARTMENT

WINDOW 2

CREW COMPARTMENT

WINDOW 1

AFT COMPARTMENT

WINDOW 3

WINDOW 4

WINDOW 5

10 FT 7 IN

ROLL ENGINES

AFT PITCH ENGINES

ABOVE The conical sidewalls provided inset windows affording ample view forward for docking with another vehicle at the apex end of the spacecraft. The top of the conical structure incorporated docking equipment and parachutes for recovery after re-entry through the atmosphere. *(North American Rockwell)*

BELOW Nomenclature for the interior layout, based on assigning the longitudinal axis through the apex of the cone back through the centre of the circular base heat shield. *(North American Rockwell)*

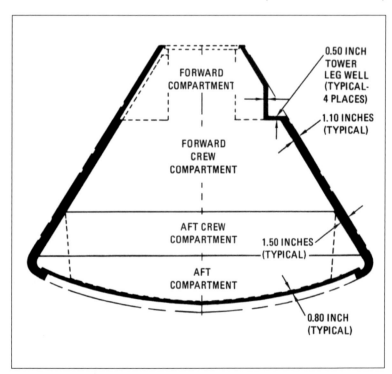

FORWARD COMPARTMENT

FORWARD CREW COMPARTMENT

AFT CREW COMPARTMENT

AFT COMPARTMENT

0.50 INCH TOWER LEG WELL (TYPICAL- 4 PLACES)

1.10 INCHES (TYPICAL)

1.50 INCHES (TYPICAL)

0.80 INCH (TYPICAL)

at the base to 0.25in (0.6cm) at the apex. The structure supported two hatches and five windows as well as ten RCS thrusters and associated propellant tanks.

The side hatch, used primarily for getting in and out of the CM on the ground, was 29in (73.7cm) high and 34in (86.4cm) wide with a 9in (22.9cm) diameter window. The CM had four other windows, two 13 x 13in (33 x 33cm) located each side of the crew compartment adjacent to the outer couches, and two triangular rendezvous windows, 8 x 13in (20.3 x 33cm) positioned as forward-facing windows for the left and right seated astronauts respectively, through which they could view along the conical walls of the module during rendezvous and docking.

The three foldable couches were identical and provided for the Commander (CDR) on the left, the Command Module Pilot (CMP) in the centre and the Lunar Module Pilot (LMP) on the right. These seat positions could change according to the phase of the mission – for instance the CMP usually occupied the left couch for rendezvous and docking so as to get a visual alignment through the window. The couches were not fixed to the base of the module but were suspended on struts with armrests and footrests that could be articulated to a wide variety of positions.

The relative positions of various locations in the module were named with respect to the couches. The base below the couches was the aft bulkhead, the upper equipment bay was the area around the ingress hatch, and the lower equipment bay was on the opposite side, beyond the footrests of the couches. The left-hand equipment bay was to the side of the left couch, the right-hand equipment bay to the side of the right couch. The upper equipment bay was at the apex of the conical module where the docking tunnel was located for access to *Aquarius*.

The habitable volume was pressurised with pure oxygen at about 5lb/in² (34.5kPa) using oxygen from the cryogenic tanks in the Service Module and controlled through the Environmental Control System (ECS) situated behind panels in the left-hand equipment bay. The average temperature was maintained at 70–75°F (21.1–23.9°C).

BELOW The Command Module provided space for three crew members, occupying foldable couches held beneath the control and displays console by struts, affording space beneath for sleeping bags. The G&N (guidance and navigation) station was situated on the lower equipment bay (LEB), where a telescope and sextant were mounted in the wall. *(North American Rockwell)*

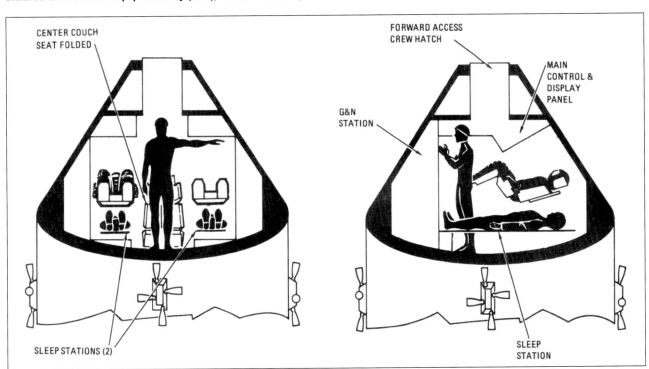

APOLLO COMMAND MODULE INTERIOR

LEFT SIDE

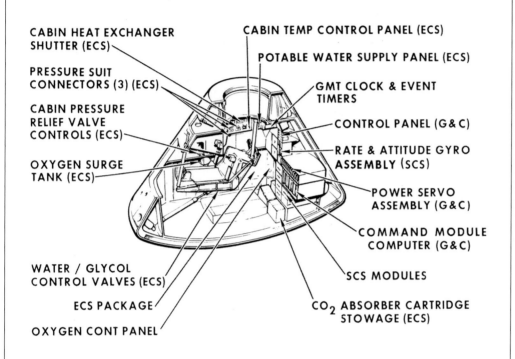

CABIN HEAT EXCHANGER SHUTTER (ECS)

PRESSURE SUIT CONNECTORS (3) (ECS)

CABIN PRESSURE RELIEF VALVE CONTROLS (ECS)

OXYGEN SURGE TANK (ECS)

WATER / GLYCOL CONTROL VALVES (ECS)

ECS PACKAGE

OXYGEN CONT PANEL

CABIN TEMP CONTROL PANEL (ECS)

POTABLE WATER SUPPLY PANEL (ECS)

GMT CLOCK & EVENT TIMERS

CONTROL PANEL (G & C)

RATE & ATTITUDE GYRO ASSEMBLY (SCS)

POWER SERVO ASSEMBLY (G & C)

COMMAND MODULE COMPUTER (G & C)

SCS MODULES

CO_2 ABSORBER CARTRIDGE STOWAGE (ECS)

RIGHT SIDE

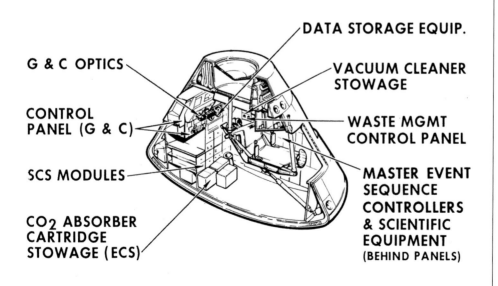

G & C OPTICS

CONTROL PANEL (G & C)

SCS MODULES

CO_2 ABSORBER CARTRIDGE STOWAGE (ECS)

DATA STORAGE EQUIP.

VACUUM CLEANER STOWAGE

WASTE MGMT CONTROL PANEL

MASTER EVENT SEQUENCE CONTROLLERS & SCIENTIFIC EQUIPMENT (BEHIND PANELS)

LEFT The Command Module contained a compact arrangement of lockers and access panels, with a large number of switches and displays. It contains almost 15 miles (24km) of electrical wiring, and two million functional parts. The CM's crew displays have 24 instruments, 566 switches, 40 mechanical event indicators and 71 lights. *(North American Rockwell)*

RIGHT The aft (base) end of the Command Module carried special collapsing struts and shock-absorbing material to allow modest deformation when landing on water. In other respects the vehicle was built for space, with a major effort made to open up the interior volume as far as possible for the crew to move around.
(North American Rockwell)

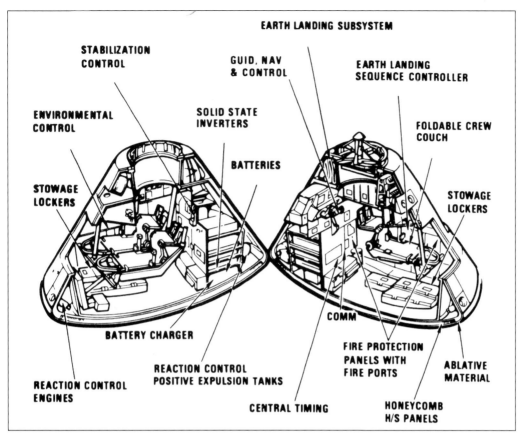

EARTH LANDING SUBSYSTEM

STABILIZATION CONTROL

GUID, NAV & CONTROL

EARTH LANDING SEQUENCE CONTROLLER

ENVIRONMENTAL CONTROL

SOLID STATE INVERTERS

FOLDABLE CREW COUCH

BATTERIES

STOWAGE LOCKERS

STOWAGE LOCKERS

BATTERY CHARGER

COMM

FIRE PROTECTION PANELS WITH FIRE PORTS

ABLATIVE MATERIAL

REACTION CONTROL POSITIVE EXPULSION TANKS

REACTION CONTROL ENGINES

CENTRAL TIMING

HONEYCOMB H/S PANELS

LEFT Interior accommodations include stowage space for equipment used infrequently or for special purposes and which could be put away when not in use. Stowage lockers located in the left and lower equipment bays provide 5,072in^3 (83,130cm^3) of storage volume for food sufficient to last three people up to 11 days at an average 2,500kcal per person per day. *(North American Rockwell)*

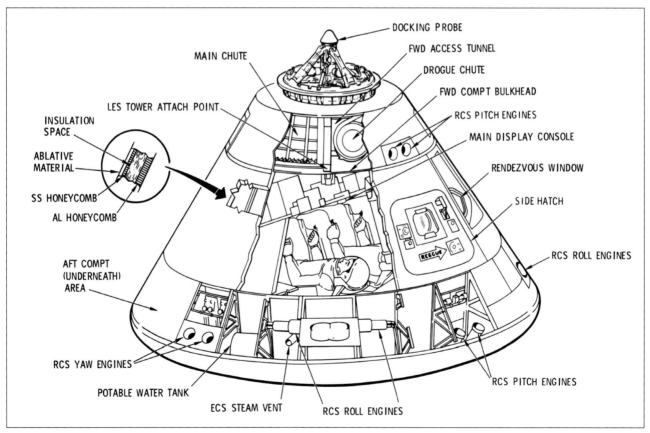

DOCKING PROBE

FWD ACCESS TUNNEL

MAIN CHUTE

DROGUE CHUTE

FWD COMPT BULKHEAD

LES TOWER ATTACH POINT

RCS PITCH ENGINES

INSULATION SPACE

MAIN DISPLAY CONSOLE

ABLATIVE MATERIAL

RENDEZVOUS WINDOW

SS HONEYCOMB

SIDE HATCH

AL HONEYCOMB

RCS ROLL ENGINES

AFT COMPT (UNDERNEATH) AREA

RCS YAW ENGINES

POTABLE WATER TANK

ECS STEAM VENT

RCS ROLL ENGINES

RCS PITCH ENGINES

LOWER EQUIPMENT BAY (LEB)

1. AUXILIARY TEST PANEL 101
2. OPTICS STOWAGE COMPARTMENT
3. LIGHTING CONTROL PANEL 101
4. OPTICS PANEL 121
5. LEB DISPLAY KEYBOARD PANEL 140
6. G&N CONTROL PANEL 122
7. B1 STOWAGE COMPARTMENT (FOOD)
8. TRANSLATION CONTROL LEB MOUNT
9. ROTATIONAL CONTROL LEB MOUNT
10. B2 STOWAGE COMPARTMENT (MED KIT)
11. B3 STOWAGE COMPARTMENT (CAMERA EQUIP)
12. B4 STOWAGE COMPARTMENT (CHLORINE EQUIP)
13. B5 STOWAGE COMPARTMENT (CO_2 ABSORBERS)
14. B6 STOWAGE COMPARTMENT (CO_2 ABSORBERS)
15. B7 STOWAGE COMPARTMENT
16. ACCESS PANEL FOR VHF TRIPLEXER
17. B8 STOWAGE COMPARTMENT (CAMERA EQUIP)

AFT BULKHEAD (AB)

1. FIRE EXTINGUISHER
2. A3 STOWAGE LOCKER
3. A4 STOWAGE LOCKER
4. A5 STOWAGE LOCKER
5. A6 STOWAGE LOCKER
6. A7 STOWAGE LOCKER
7. ELECTRICAL CABLEWAY
8. A8 STOWAGE LOCKER

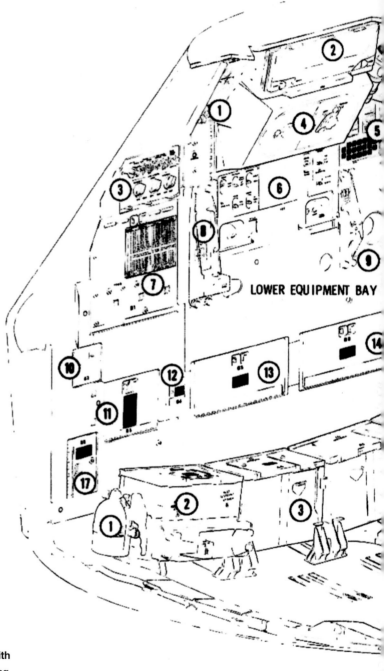

A summary of the copious locker and storage provision with some stowed items identified. With the 'long' axis extending from the base of the heat shield up through the centre of the conical module, the 'lower equipment bay' was the wall that lay beyond the feet of the astronauts in their couches, 'left' and 'right' equipment bays falling to the conventional left and right sides of the spacecraft. *(North American Rockwell)*

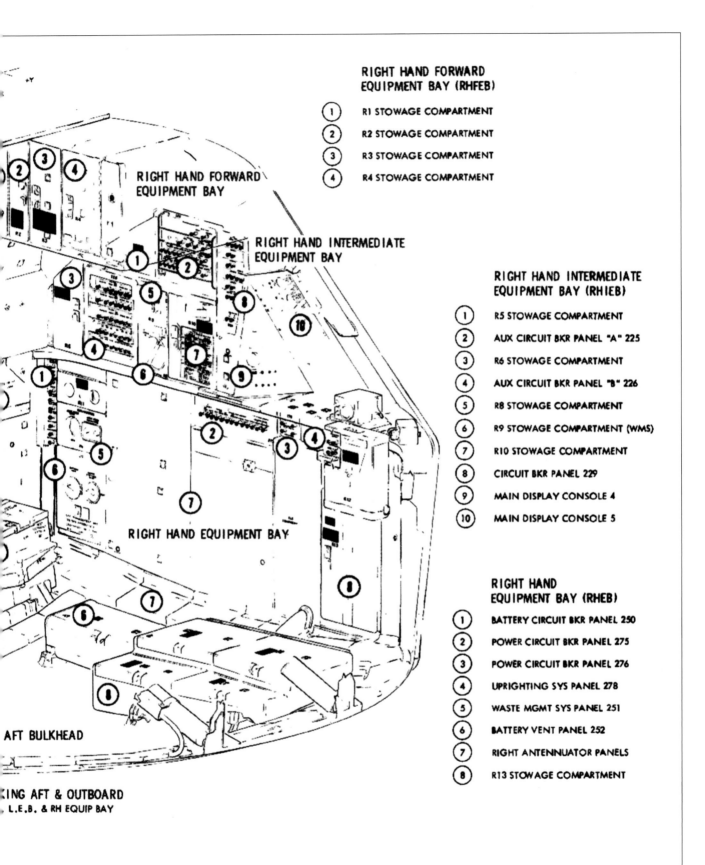

RIGHT HAND FORWARD
EQUIPMENT BAY (RHFEB)

1. R1 STOWAGE COMPARTMENT
2. R2 STOWAGE COMPARTMENT
3. R3 STOWAGE COMPARTMENT
4. R4 STOWAGE COMPARTMENT

RIGHT HAND FORWARD
EQUIPMENT BAY

RIGHT HAND INTERMEDIATE
EQUIPMENT BAY

RIGHT HAND INTERMEDIATE
EQUIPMENT BAY (RHIEB)

1. R5 STOWAGE COMPARTMENT
2. AUX CIRCUIT BKR PANEL "A" 225
3. R6 STOWAGE COMPARTMENT
4. AUX CIRCUIT BKR PANEL "B" 226
5. R8 STOWAGE COMPARTMENT
6. R9 STOWAGE COMPARTMENT (WMS)
7. R10 STOWAGE COMPARTMENT
8. CIRCUIT BKR PANEL 229
9. MAIN DISPLAY CONSOLE 4
10. MAIN DISPLAY CONSOLE 5

RIGHT HAND
EQUIPMENT BAY (RHEB)

1. BATTERY CIRCUIT BKR PANEL 250
2. POWER CIRCUIT BKR PANEL 275
3. POWER CIRCUIT BKR PANEL 276
4. UPRIGHTING SYS PANEL 278
5. WASTE MGMT SYS PANEL 251
6. BATTERY VENT PANEL 252
7. RIGHT ANTENNUATOR PANELS
8. R13 STOWAGE COMPARTMENT

RIGHT HAND EQUIPMENT BAY

AFT BULKHEAD

:ING AFT & OUTBOARD
L.E.B. & RH EQUIP BAY

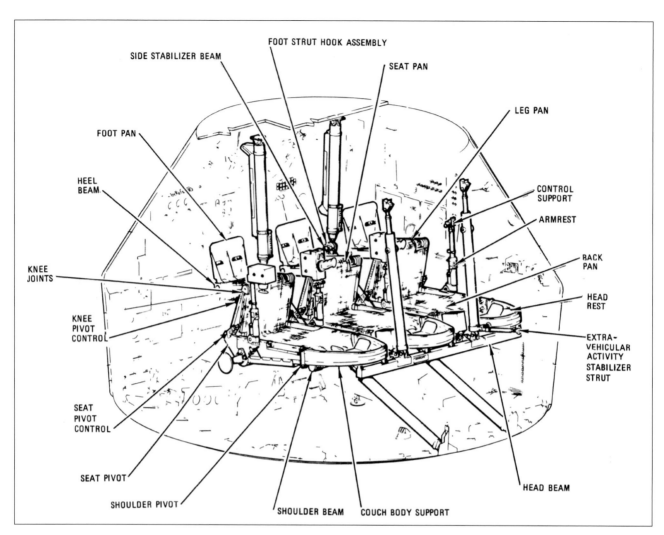

FOOT STRUT HOOK ASSEMBLY

SIDE STABILIZER BEAM

SEAT PAN

LEG PAN

FOOT PAN

CONTROL SUPPORT

ARMREST

HEEL BEAM

BACK PAN

KNEE JOINTS

HEAD REST

KNEE PIVOT CONTROL

EXTRA-VEHICULAR ACTIVITY STABILIZER STRUT

SEAT PIVOT CONTROL

SEAT PIVOT

HEAD BEAM

SHOULDER PIVOT

SHOULDER BEAM   COUCH BODY SUPPORT

ABOVE The couches are standard size for all missions, each capable of being folded, and made of hollow steel tubing covered with Armalon, a fibreglass cloth. They rest on a head beam and two side stabiliser beams supported by eight attenuator struts, two each for the Y and Z axes and four for the X axis, which absorb the impact of landing. *(North American Rockwell)*

RIGHT Looking across the three couches, the lower equipment bay supports the guidance and navigation station. *(North American Rockwell)*

LEFT A view through the main access hatch showing the centre couch usually occupied by the Command Module Pilot, and the right couch for the Lunar Module Pilot. The couch restraint system consists of a lap belt and two shoulder straps which connect together at a buckle. The shoulder straps connect to the shoulder beam of the couch. *(North American Rockwell)*

BELOW A rotation controller for changing the attitude orientation of the CSM is located here on each of the two inner armrests on opposing couches. Note the display and keyboard assembly (DSKY) at upper left. *(North American Rockwell)*

BELOW Electronic circuits being installed behind panel 1 of the main display console. *(North American Rockwell)*

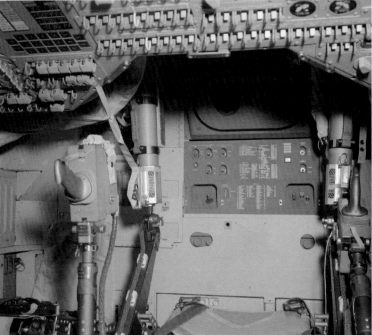

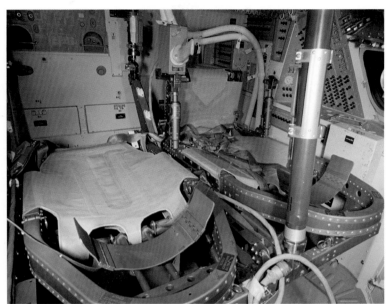

ABOVE A view looking from the left couch area across to the right-hand equipment bay (at the right) and the lower equipment bay, where the location of the two optical devices can be seen. Also visible is the support structure for the controls and displays console which is installed directly forward of (above) all three couches. *(North American Rockwell)*

ABOVE Approximately the same orientation as the picture to the left but with equipment fully installed for flight. *(North American Rockwell)*

BELOW Looking right across the couches, the Lunar Module Pilot's couch is placed in the 85° launch and landing position. Two rotation hand controllers are attached to the two inner armrests and a good view is afforded of the main access hatch and its locking handle. *(North American Rockwell)*

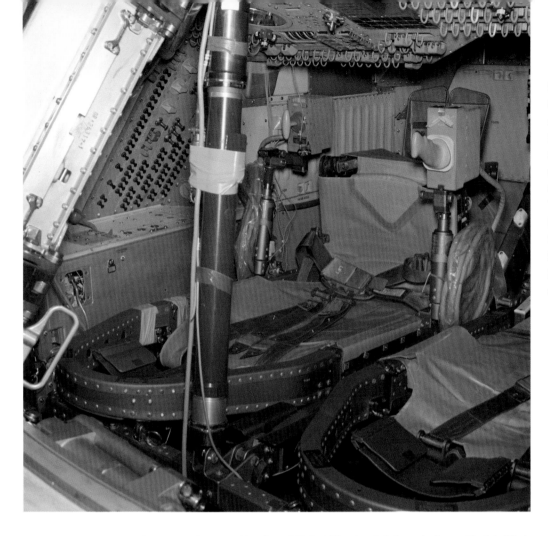

LEFT The view across to the left side of the Command Module with the environmental control system on the left side of the spacecraft. The headrests can be adjusted 6.5in (16.5cm) up or down for individual crew preference. *(North American Rockwell)*

BELOW Circuit breaker panel No 8 with white circular lighting switches. *(North American Rockwell)*

BELOW The view left from the Lunar Module Pilot's couch shows the location of windows 1 (side), 2 (forward for rendezvous and docking) and 3 (centre hatch). The headrests have hinged side flaps. *(North American Rockwell)*

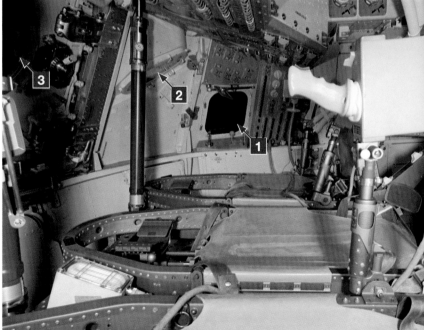

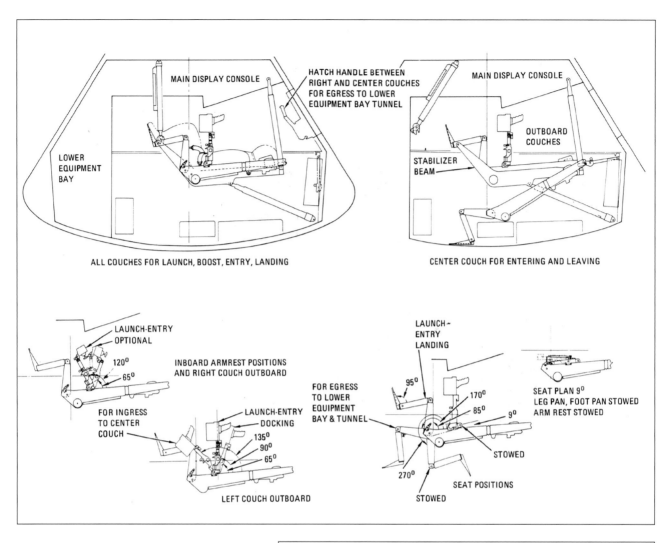

MAIN DISPLAY CONSOLE

HATCH HANDLE BETWEEN RIGHT AND CENTER COUCHES FOR EGRESS TO LOWER EQUIPMENT BAY TUNNEL

LOWER EQUIPMENT BAY

ALL COUCHES FOR LAUNCH, BOOST, ENTRY, LANDING

MAIN DISPLAY CONSOLE

OUTBOARD COUCHES

STABILIZER BEAM

CENTER COUCH FOR ENTERING AND LEAVING

LAUNCH-ENTRY OPTIONAL

120°
65°

INBOARD ARMREST POSITIONS AND RIGHT COUCH OUTBOARD

FOR INGRESS TO CENTER COUCH

LAUNCH-ENTRY DOCKING
135°
90°
65°

LEFT COUCH OUTBOARD

LAUNCH-ENTRY LANDING

FOR EGRESS TO LOWER EQUIPMENT BAY & TUNNEL

95°
170°
85°
9°

270°

STOWED

SEAT POSITIONS

STOWED

SEAT PLAN 9°
LEG PAN, FOOT PAN STOWED
ARM REST STOWED

ABOVE The different couch positions provide wide variation according to the phase of the mission or the flight situation. Two armrests are attached to the back pan for the left and right couches, with no armrests on the centre couch. Translation and rotation hand controllers can be attached to any one of the armrests. The couch seat pan and leg pan are formed from framing and cloth, while the foot pan is solid steel.
(North American Rockwell)

RIGHT Sleeping bags are slung beneath the outer couches. Fabricated from lightweight Beta cloth, they are 64in 162.6cm) long with zipper openings for the torso and 7in (17.8cm) neck openings.
(North American Rockwell)

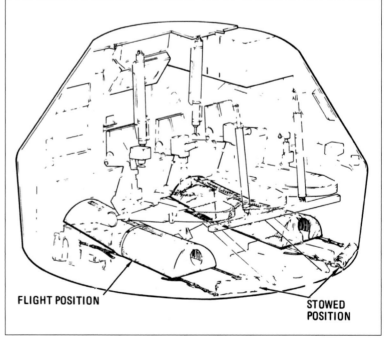

FLIGHT POSITION

STOWED POSITION

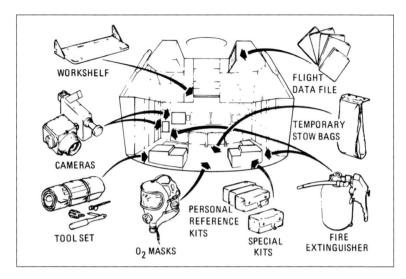

LEFT Operational aids are contained in various places, and include a comprehensive flight data file situated in the right-hand forward equipment bay, tool set, oxygen masks for use when the cabin air is contaminated, personal preference kits (PPKs) containing privately selected items, and a fire extinguisher on the aft equipment bay. *(North American Rockwell)*

LEFT Technicians working on the interior assembly of a Command Module take meticulous care over tools and sundry items that could cause damage if left on board for flight. The hatch is secured with 12 latches. *(North American Rockwell)*

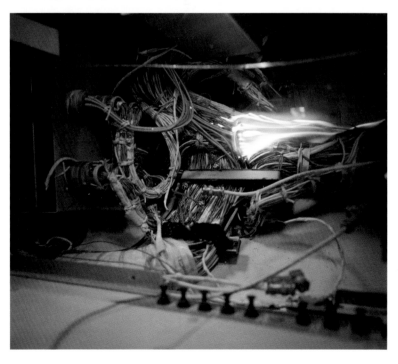

BOTTOM Electrical wiring in the CSM was manufactured to a common standard, insulation being capable of withstanding 1,500vdc with conductors tested to perform to a temperature of up to 500°F (260°C) without compromising the wiring or its protective coating. Teflon insulation 15 mils (0.015in or 0.381mm) thick was applied as standard to early (unmanned) Block I spacecraft to protect against abrasions and damage during manufacture and installation of harnesses. Most conductors were nickel-plated copper and the thinest wire used was 24-gauge. The Block I CSM contained 110,000ft (33,528m) of wiring with a total weight of 1,850lb (839kg), but Teflon insulation in Block II spacecraft for manned missions was 7 mils thick with 0.5 mil polymide dispersion coating for added safety, saving 500lb (226.8kg) in weight. Materials used in the spacecraft limited any fire to a maximum induced temperature of 400°F before it could be extinguished, well below the wiring insulation limit. Circuit breakers and wire sizes were modified to maintain circuit integrity. A major problem emerged where the polymide dispersion coating on insulation showed cracking due to improper curing, and a considerable amount of wiring had to be replaced to correct this. *(NASA)*

The Lunar Module

The LM comprised two stages: the pressurised Ascent Stage for housing the two-man crew, with its own Ascent Propulsion System (APS) for ascending back into Moon orbit, and the unpressurised Descent Stage with supporting legs and a Descent Propulsion System (DPS) for controlled descent and landing.

A significant decision, highly relevant to Apollo 13, was to plan for use of the LM as a 'lifeboat' in the event that the CSM became disabled en route to the Moon. That capability arose out of the Apollo Mission Planning Task Force (AMPTF) set up in January 1964 under Tom Barnes at the instigation of Joseph Shea following a suggestion by Grumman's Thomas J. Kelly.

In the ensuing analysis the inherent value in using the LM as a contingency lifeboat was accommodated by expanding the volume of the Descent Stage propellant tanks 10–15% above that needed for the design reference mission (the DRM). Instead of comprising spherical tanks 51in (129.5cm) in diameter, each of the four fuel and oxidiser tanks in the Descent Stage would be extended to a length of 70.8in (180cm) by way of a cylindrical girth between the two hemispherical end domes, increasing the capacity and extending the delta velocity which might under extreme circumstances be needed to get the CSM home.

In reference to the axial geometry of the Lunar Module, the Z-axis ran through the vehicle front to back for roll; +Z facing forwards through the front hatch, -Z facing aft. The lateral Y-axis ran at right angles to the Z-axis for pitch manoeuvres, +Y pointing to the right, -Y to the left. The vertical X-axis for yaw had +X pointing up through the docking hatch, -X pointing down through the Descent Stage engine bell.

The Ascent Stage

The Ascent Stage comprised three sections: crew compartment, midsection and aft equipment bay. The underlying structure took the form of a barrel in the horizontal plane, the front end being the crew compartment where the astronaut control stations were located, the centre being the midsection, into the floor of which the Ascent Engine Assembly (AEA) was located. The top of the midsection housed the upper hatch and docking collar. The unpressurised aft equipment bay was attached to the outer section of the bulkhead at the rear of the pressurised section.

The structural shell of the cylindrical crew compartment was of semi-monocoque aluminium alloy, chemically-milled skins and machined longerons. The shell was supported by formed Z-sheet metal rings riveted to the structural skin to form a barrel 7.67ft (2.34m) in diameter and 3.5ft (1.07m) in length. The front face of the crew compartment had openings for the forward hatch, 32in (81.3cm) square, and two triangular forward-facing windows, each with an area of 2ft^2 (0.19m^2), canted downwards to provide lateral and downward-facing views. The docking window above the Commander's station had an area of 80in^2 (516cm^2).

The Commander's flight station was at the left, and the LM pilot's station at the right, with station centrelines 44in (112cm) apart and controls and display panels up front between the crew members' standing positions and the two canted windows. Additional switches and circuit breakers were situated on respective sidewalls and there were side translation and attitude hand controllers for each crew member. In reality the Commander 'flew' the LM and the LM pilot was more a systems engineer, calling out instrument readings and computer display information while the CDR had his eyes out the window.

The midsection consisted of a ring-stiffened, semi-monocoque shell similar to that of the crew compartment, with fusion-welded longerons and machined stiffeners. It was strengthened at the lower section for the Ascent Engine Assembly. The upper section was strengthened to accommodate the docking ring and the 33in (83.8cm) diameter overhead hatch

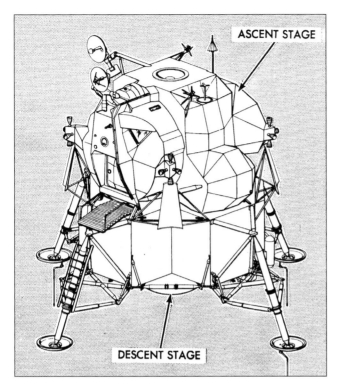

ASCENT STAGE

DESCENT STAGE

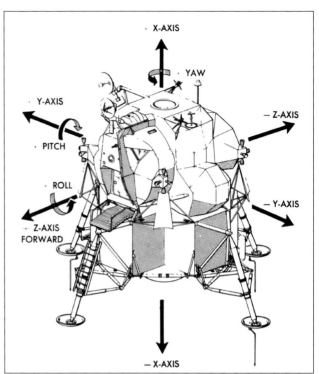

X-AXIS

YAW

Y-AXIS

Z-AXIS

PITCH

ROLL

Y-AXIS

Z-AXIS
FORWARD

X-AXIS

ABOVE The Lunar Module was developed for carrying two astronauts down to the surface of the Moon, sustaining 33 hours of surface operations and delivering them back to the orbiting Apollo spacecraft. Comprising a Descent Stage accommodating the descent propulsion system and rocket motor, and an Ascent Stage for supporting two crew members in a habitable environment, the LM was sized so as to be capable of performing docked propulsive manoeuvres should the Apollo spacecraft be disabled. *(Grumman)*

ABOVE The three-axis coordinate system on the LM provided an orientation through which guidance, navigation and control functions were conducted for all attitude and translation manoeuvres. *(NASA)*

BELOW LEFT Lunar Test Article No 1 (LTA-1) being prepared for checks in the Electromagnetic Interference Room at Grumman in Bethpage, New York. *(Grumman)*

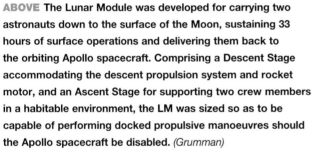

LEFT The front face of LM-5 (Apollo 11) displays the square forward egress hatch affording access to the porch and ladder on the Descent Stage. Note the asymmetric appearance due to the offset Aerozene-50 fuel tank for the Ascent Propulsion System mounted on the -Y side of the structure. *(Grumman)*

incorporating the docking tunnel, 18in (45.7cm) in length, leading directly to the Command Module when docked.

There was an 18in (45.7cm) step up from the crew compartment to the midsection, which in weightlessness or one-sixth gravity on the Moon posed no problem. The midsection was 4.5ft (1.37m) in length and about 5ft (1.52m) high, but its shaped 'walls' gave it a width of no more than 4.7ft (1.43m).

As a single open area the pressurised crew compartment and midsection were 8ft (2.44m) in length from front to back, with an internal volume of 235ft³ (6.65m³). In theory this gave each crew member more than 40% greater space than in the Command Module, but the interior shape of the Ascent Stage compromised that potential. There were stowage spaces for the helmet bags, the drogue removed from the docking tunnel, the backpacks and numerous small items, including flight data files.

Supported on outriggers attached to the midsection, four quads of Reaction Control System (RCS) thrusters provided attitude control and translation manoeuvres throughout independent flight. Propellant was contained in four cylindrical tanks, two either side of the midsection, one for fuel and the other for oxidiser. The two parallel systems, A and B, were independent but with a propellant cross-feed that enabled the crew to use both or either. Each tank contained a Teflon bladder within which the propellant was contained. Helium gas was supplied to the area between the tank wall and the bladder to expel the contents.

The four thruster quads were situated on four arms extending outward from the four corners of the Ascent Stage for attitude control and stabilisation. They could also be used for translation and orbital changes during ascent from the Moon and rendezvous and docking with the Apollo spacecraft. Each quad carried four rocket motors, each with a nominal thrust of 100lb (448N) and a chamber pressure of 96lb/in² (662kPa). They operated on Aerozene-50/nitrogen tetroxide at a fuel/oxidiser ratio of 1:2.05. Each thruster was about 13.5in (34.3cm) long, designed with a restart capability of up to 10,000 times.

The two system A propellant tanks on the -Y side of the midsection included an oxidiser tank

forward and a fuel tank aft. Both tanks were 12.6in (32cm) in diameter, fabricated from 6A1-4V titanium. The oxidiser tank had a length of 38.2in (97cm) with a volume of 2.38ft³ (0.07m³), containing 208.8lb (94.71kg) of nitrogen tetroxide. The fuel tank had a length of 32.2in (81.8cm), a volume of 1.91ft³ (0.005m³), and contained 107.7lb (48.85kg) of Aerozene-50.

Between the two propellant tanks at the bottom was a single spherical tank, 12.3in (31.2cm) in diameter with a volume of 910in³ (14,915cm³) containing 1.03lb (0.47kg) of helium to pressurise the RCS propellant tanks for system A. System B consisted of an identical set of propellant and helium tanks situated on the opposite (+Y) side of the midsection but with the position of the propellant tanks reversed, the fuel tank being forward of the oxidiser tank.

Two water tanks were situated externally either side of the top midsection, above the RCS propellant tanks. Fabricated from 6061-T6 aluminium, the water tanks had a diameter of 14.5in (38.8cm) and each contained 42.1lb (19.1kg) of water. Initially charged with nitrogen to 48.2lb/in² (332.3kPa), the water was routed to a control module which incorporated a valve for selecting between Ascent or Descent water supply. These two Ascent Stage tanks held only 25% of water available from both Stages, the Ascent quantity supplying water for cooling, drinking and food preparation after lift-off from the Moon.

Power for getting off the Moon or for aborting during descent to the surface, came from the Ascent Propulsion System (APS) and its Ascent Engine Assembly (AEA). With a nominal fixed thrust of 3,500lb (15.57kN) the engine could be fired up to 35 times, but due to the materials utilised in its assembly it could not be used more than 40 hours after first ignition. The engine operated on a fuel/oxidiser ratio of 1:1.6, with ablative cooling in the combustion chamber and expansion skirt, or nozzle. It weighed 180lb (81.6kg), with a height of 47in (119.4cm) and a nozzle skirt diameter of 34in (86.4cm). Combustion chamber pressure was 170lb/in² (1,172kPa). Installed within the floor of the midsection, it was tilted 1.5° off the X-axis in the +Z direction.

Aerozene-50 and nitrogen tetroxide propellant for the APS was contained in identically sized

ABOVE The +Y side of the LM where the N_2O_4 oxidiser tank for the Ascent Propulsion System is located close in to the main pressure cabin. *(Grumman)*

BELOW A three-quarter rear port view of the Ascent Stage with the aft equipment bay and the two VHF antennae on opposing positions above the cabin midsection. *(Grumman)*

BELOW RIGHT Starboard side view of the Ascent Stage with the alloy N_2O_4 oxidiser tank. One of two black-coloured helium pressurisation tanks is clearly visible behind the mid-section and forward of the aft equipment bay. *(Grumman)*

ABOVE Viewed from the three-quarter rear starboard side, the aft equipment bay is set on a series of outriggers attached to the aft mid-section bulkhead, and carries electronic equipment, electrical devices and two batteries. The black dish-shaped antenna provides high-gain S-band communications with the Earth. The white circular rear-facing dish on the forward face of the Ascent Stage is the rendezvous radar antenna, not used on Apollo 13. This view shows clearly the stand-off frame used to support the external thermal insulation. *(NASA-KSC)*

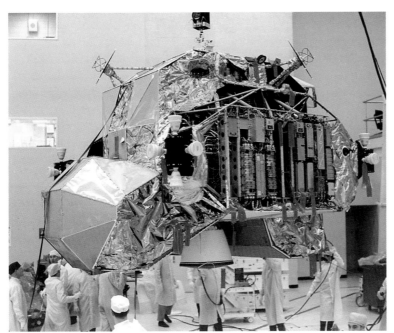

tanks located on either side of the midsection. Each tank was fabricated from 6A1-4V titanium and had a diameter of 49.4in (125.5cm) with an internal volume of 36ft³ (1.02m³). Because of the different molecular mass of the propellants, the tank on the -Y side of the Ascent Stage held 2,009lb (911.3kg) of fuel while the +Y tank held 3,220lb (1,460kg) of oxidiser.

Each tank was mounted on outriggers, but the lighter fuel tank was held farther out from the geometric centre of the Ascent Stage so that it would compensate for the difference in mass, keeping the Stage in balance but giving it an asymmetric appearance.

At the bottom of the midsection aft bulkhead, on opposite sides of the structure, were two identical helium pressurisation tanks for the Ascent Propulsion System, one serving the APS fuel tank, the other the oxidiser tank. Each tank was fabricated from 6A1-4V titanium, with a diameter of 22.3in (56.6cm) and a volume of 3.35ft³ (0.09m³), and contained 6.6lb (2.99kg) of helium. Gaseous helium held at 3,050–3,500lb/in² (21,030–24,132kPa) pressurised the propellant tanks for delivery of the fuel and the oxidiser to the main engine.

The aft equipment bay was an unpressurised area attached to the exterior of the aft midsection bulkhead and was cantilevered approximately 33in (83.8cm) out of the rear of the Ascent Stage. It was here that most of the electronic controls and electrical systems were situated supporting communications, electrical production and guidance and navigation systems. The banks of modules and electronic equipment were attached to the frame that enclosed the two helium pressurisation tanks and the two oxygen tanks.

Of the six batteries in the LM, two were in the aft equipment bay and four were in the Descent Stage. The two Ascent Stage silver-zinc batteries each had 20 cells with potassium hydroxide as the electrolyte. At a nominal 30V DC, each battery weighed 125lb (56.7kg) and had a capacity of 296Ah. Each was 35.25in (89.5cm) in length with a width of 8.1in (20.6cm) and a depth of 6in (15.2cm). These batteries were not normally used in parallel, providing two independent systems for the electrical requirements of the stage after separation from the Descent Stage prior to lift-off from the Moon.

The aft equipment bay carried two identical gaseous oxygen tanks, 11.97in (30.4cm) in diameter, fabricated from Inconel 718 and mounted to the outer shell above the two helium tanks and the helium pressurisation control modules. Each tank held approximately 2.4lb (1.09kg) of oxygen, with a nominal pressure of 854lb/in² (5,888kPa).

Excess heat produced by electrical equipment in the aft equipment bay, and in the pressurised part of the spacecraft, was removed by cold plates and eight cold rails placed vertically across the bay. A solution of 35% ethylene glycol and 65% water was used in the 35lb (15.9kg) of coolant carried for the heat transport system (HTS), which consisted of a primary and a secondary loop. The purpose of the HTS was to remove heat from electronic equipment and the batteries in both Ascent and Descent stages, and to cool the cabin atmosphere, the suit circuits and electronic equipment and modules.

Using a glycol pump, coolant was forced through the system, which used a water sublimator to reject heat to space and, by venting, generated steam through a special non-propulsive vent. The primary loop alone was responsible for using excess heat to warm the pressurised section of the Ascent Stage. The water-glycol coolant loop was charged with the solution at the fill point and sealed for the duration of the mission. The glycol accumulator maintained a head pressure of 5.25–9lb/in² (36.2–62.1kPa) in the system. Coolant from the recirculation assembly served cold plates in both pressurised and unpressurised areas and could be redirected through both rails and plates for maximum heat transport.

The Ascent Stage also provided a mounting for the rendezvous radar and for the high-gain S-band antenna. The 24in (61cm) parabolic radar antenna was mounted to the upper forward section of the midsection on a U-mount projecting forward from the front face of the Ascent Stage. It had a transmit frequency at 9832.8MHz with a receive frequency of 9792.0MHz +/- Doppler. The return signal was received by a four-point feed-horn attached to four struts at the focus of the Cassegrain antenna.

Communications between the LM and the CSM was by VHF through either one of two

ABOVE The pressurised mid-section and crew compartment display the strong structural elements on to which are attached support structures and systems. The Aerozene-50 fuel tank carries a work cover, directly above which are the two black RCS propellant tanks, fuel to the right, oxidiser to the left. Between them, immediately below the hoist, is the connector panel for the helium-pressurisation module. Also clearly visible between the aft mid-section and the cantilevered aft equipment bay is one of two helium pressurisation tanks. White work covers shroud cantilevered RCS-thruster quads on outriggers. *(Grumman)*

ABOVE The aft view of the Ascent Stage displays outrigger-strut geometry emphasising the different approach to spacecraft design based on a configuration free of the need to protect the structure from atmospheric friction. *(Grumman)*

BELOW Primary structural elements of the Ascent Stage, also showing Thruster Control Assemblies on outriggers. *(Grumman)*

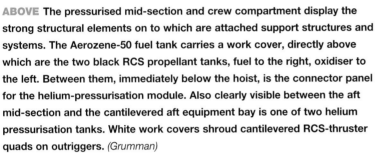

BELOW An informative view of the top of the Ascent Stage is afforded by this tilted stage displaying a water tank and the upper docking hatch. *(Grumman)*

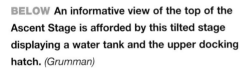

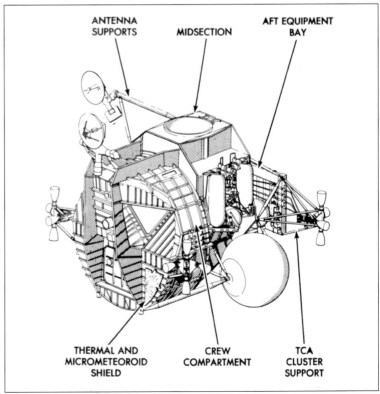

The front face of the Ascent Stage is mechanically assembled from ten welded and machined sections. After a sealing and curing operation, the outer flange contour is machined for mating to the cabin skin assembly, which is built up from chemically milled skin panels welded and mechanically fastened. The joints are then sealed and trimmed to the forward edge to match the front face contour, with added longerons and stringers to complete the assembly. The mid-section is essentially two machined bulkheads, an upper tunnel weldment, lower engine deck weldment, and chemically milled skins. This section is mechanically joined to the front face and cabin skin and sealed to form the cabin pressure shell. (Grumman)

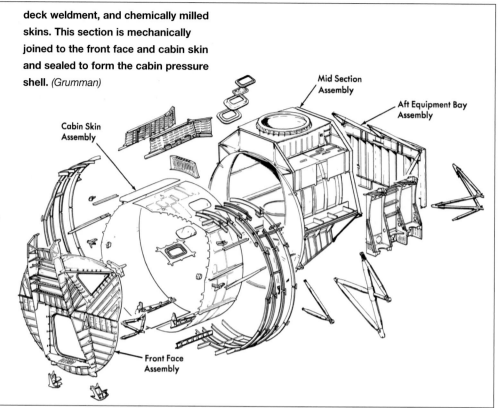

Mid Section Assembly

Aft Equipment Bay Assembly

Cabin Skin Assembly

Front Face Assembly

The Ascent Stage has approximately 20 major electrical harnesses and 60 electrical-cable assemblies. The external harnesses shown here complement internal harnesses unique to the pressure vessel. (Grumman)

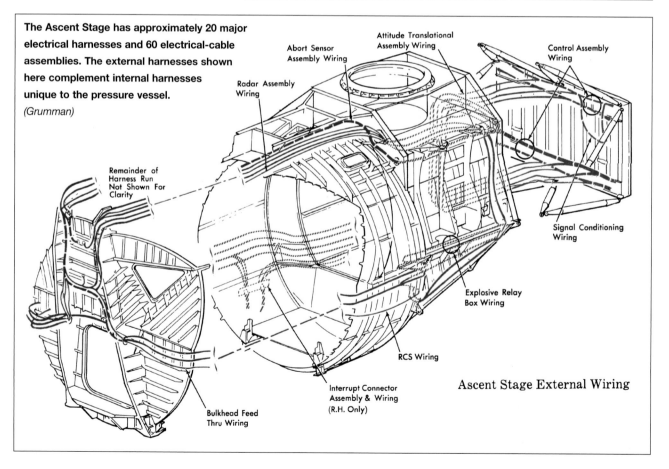

Attitude Translational Assembly Wiring

Abort Sensor Assembly Wiring

Control Assembly Wiring

Radar Assembly Wiring

Remainder of Harness Run Not Shown For Clarity

Signal Conditioning Wiring

Explosive Relay Box Wiring

RCS Wiring

Interrupt Connector Assembly & Wiring (R.H. Only)

Bulkhead Feed Thru Wiring

Ascent Stage External Wiring

ABOVE Astronaut Frank Borman gives scale to the circular cabin structure of the Ascent Stage with conduits, piping and electrical harnesses. (Grumman)

BELOW The crew cabin assembly of the Ascent Stage looking forward from the mid-section, showing primary equipment installed for crew support. (Grumman)

ABOVE The Commander's station in the crew compartment of the Ascent Stage. (Grumman)

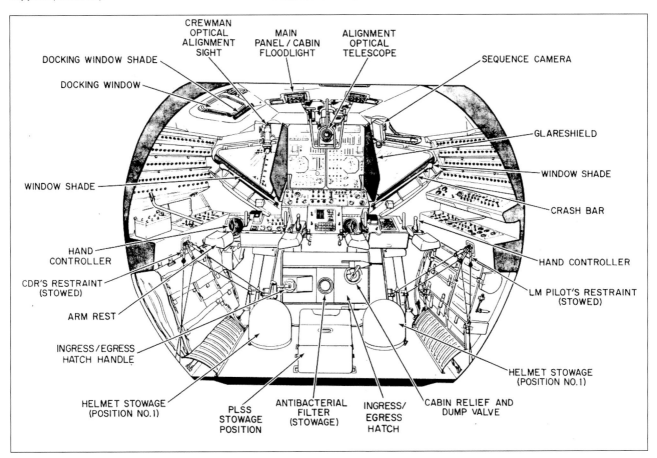

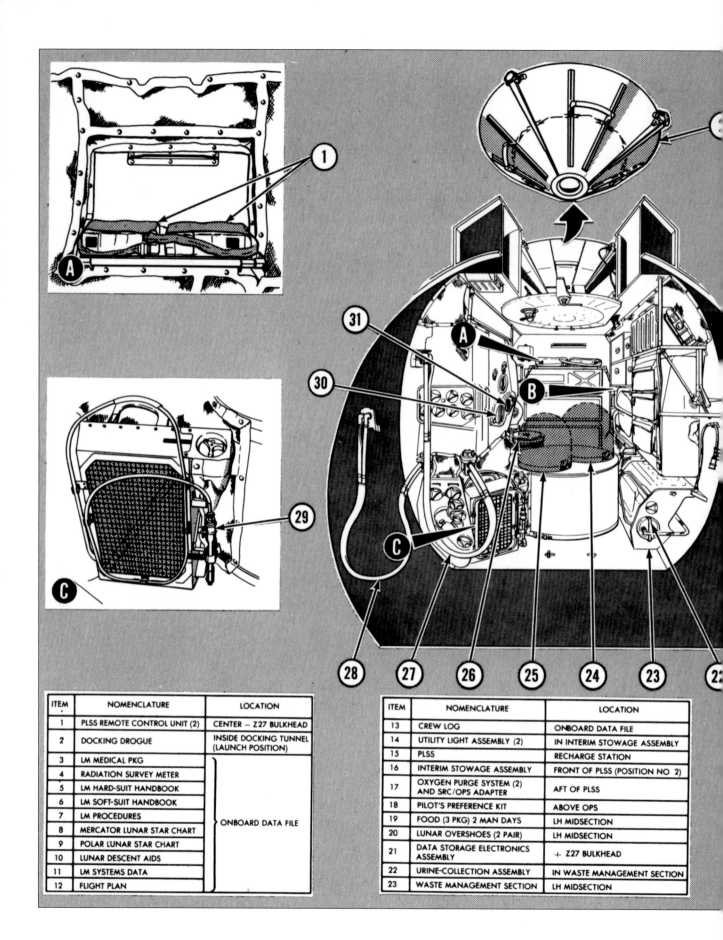

ITEM	NOMENCLATURE	LOCATION
1	PLSS REMOTE CONTROL UNIT (2)	CENTER – Z27 BULKHEAD
2	DOCKING DROGUE	INSIDE DOCKING TUNNEL (LAUNCH POSITION)
3	LM MEDICAL PKG	
4	RADIATION SURVEY METER	
5	LM HARD-SUIT HANDBOOK	
6	LM SOFT-SUIT HANDBOOK	
7	LM PROCEDURES	
8	MERCATOR LUNAR STAR CHART	ONBOARD DATA FILE
9	POLAR LUNAR STAR CHART	
10	LUNAR DESCENT AIDS	
11	LM SYSTEMS DATA	
12	FLIGHT PLAN	

ITEM	NOMENCLATURE	LOCATION
13	CREW LOG	ONBOARD DATA FILE
14	UTILITY LIGHT ASSEMBLY (2)	IN INTERIM STOWAGE ASSEMBLY
15	PLSS	RECHARGE STATION
16	INTERIM STOWAGE ASSEMBLY	FRONT OF PLSS (POSITION NO. 2)
17	OXYGEN PURGE SYSTEM (2) AND SRC/OPS ADAPTER	AFT OF PLSS
18	PILOT'S PREFERENCE KIT	ABOVE OPS
19	FOOD (3 PKG) 2 MAN DAYS	LH MIDSECTION
20	LUNAR OVERSHOES (2 PAIR)	LH MIDSECTION
21	DATA STORAGE ELECTRONICS ASSEMBLY	+ Z27 BULKHEAD
22	URINE-COLLECTION ASSEMBLY	IN WASTE MANAGEMENT SECTION
23	WASTE MANAGEMENT SECTION	LH MIDSECTION

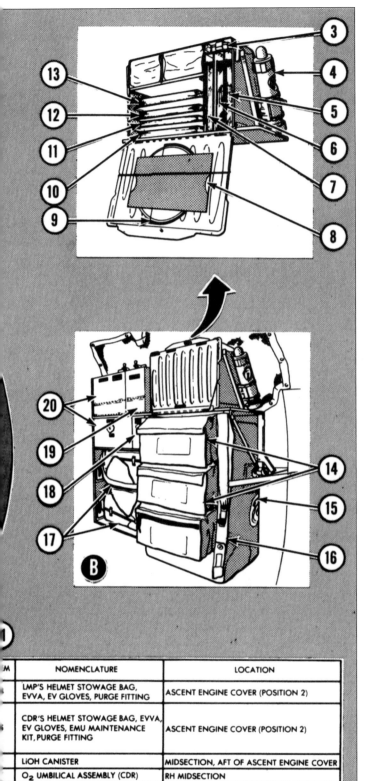

ABOVE The mid-section serves as a mounting for the Ascent Engine seen here at the bottom right without the cover it has in flight. To the left is the environmental control system equipment, and to the back of the midsection is where the backpacks are serviced for surface operations. *(Grumman)*

LEFT Looking aft, the mid-section, with a step up from the forward crew compartment and the central Ascent Engine cover supporting spacecraft systems and equipment. Also shown is the top hatch and the docking drogue installed in the tunnel for attaching the LM to the Apollo spacecraft. *(Grumman)*

RIGHT The aft section of the pressurised crew compartment where astronauts would have serviced their life-support backpacks, or PLSS (Personal Life Support System). Note the Ascent Engine cover and the upper hatch. *(Grumman)*

NOMENCLATURE	LOCATION
LMP'S HELMET STOWAGE BAG, EVVA, EV GLOVES, PURGE FITTING	ASCENT ENGINE COVER (POSITION 2)
CDR'S HELMET STOWAGE BAG, EVVA, EV GLOVES, EMU MAINTENANCE KIT, PURGE FITTING	ASCENT ENGINE COVER (POSITION 2)
LiOH CANISTER	MIDSECTION, AFT OF ASCENT ENGINE COVER
O$_2$ UMBILICAL ASSEMBLY (CDR)	RH MIDSECTION
O$_2$ UMBILICAL ASSEMBLY (LMP)	RH SIDE
WATER DISPENSER FIRE-EXTINGUISHING ASSEMBLY	FORWARD OF ENGINE COVER
LiOH CARTRIDGE (ECS)	RH AFT MIDSECTION
LiOH CARTRIDGE (PLSS)	RH AFT MIDSECTION

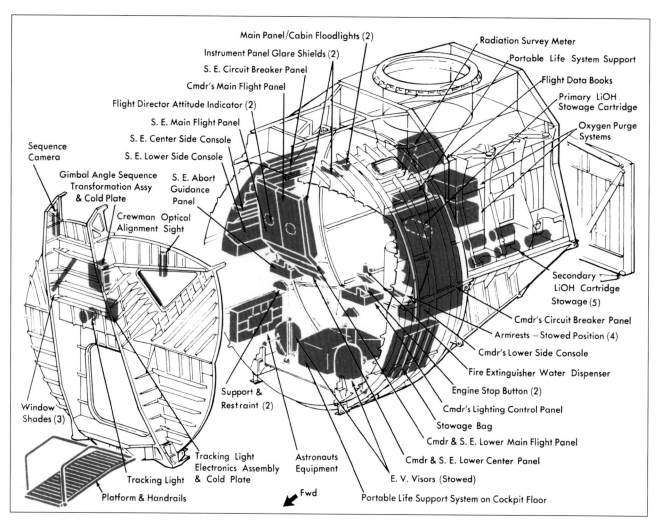

Main Panel/Cabin Floodlights (2)
Instrument Panel Glare Shields (2)
S. E. Circuit Breaker Panel
Cmdr's Main Flight Panel
Flight Director Attitude Indicator (2)
S. E. Main Flight Panel
S. E. Center Side Console
S. E. Lower Side Console
S. E. Abort Guidance Panel
Sequence Camera
Gimbal Angle Sequence Transformation Assy & Cold Plate
Crewman Optical Alignment Sight

Radiation Survey Meter
Portable Life System Support
Flight Data Books
Primary LiOH Stowage Cartridge
Oxygen Purge Systems
Secondary LiOH Cartridge Stowage (5)
Cmdr's Circuit Breaker Panel
Armrests — Stowed Position (4)
Cmdr's Lower Side Console
Fire Extinguisher Water Dispenser
Engine Stop Button (2)
Cmdr's Lighting Control Panel
Stowage Bag
Cmdr & S. E. Lower Main Flight Panel
Cmdr & S. E. Lower Center Panel
E. V. Visors (Stowed)
Portable Life Support System on Cockpit Floor

Support & Restraint (2)
Window Shades (3)
Tracking Light Electronics Assembly & Cold Plate
Tracking Light
Platform & Handrails
Astronauts Equipment
Fwd

ABOVE Crew provisions and displays are located in various positions around the interior of the pressurised cabin. Note the location of the lithium hydroxide (LiOH) cartridges, which would be a key feature in the survival story for the crew of Apollo 13. *(Grumman)*

omnidirectional, circularly polarised antennae. One was attached to struts at the top of the midsection at the rear on the +Y side facing aft, the other to struts on the upper -Y side of the midsection facing outboard. A switch on the display panel controlled which antenna was selected. The communications assembly provided two solid-stage superheterodyne receivers and two transmitters. One transmitter-receiver combination provided a 296.8MHz signal for channel A at 30/1.2W; the other a 259.7MHz signal for channel B simplex or duplex voice communications at 31.7/1.2W.

For communications with Earth in deep space and in the vicinity of the Moon, something more than the VHF and UHF systems used by previous manned spacecraft was necessary. Apollo adopted a new and revolutionary system for combining tracking and ranging together with voice, TV and data. It was known as the Unified S-band system (USB) and pioneered an

entirely new way of communication, modulating on to a carrier wave a wide range of applications within a fixed frequency. It was based on a coherent Doppler and pseudo-random range system which had already been developed by NASA's Jet Propulsion Laboratory, the home of planetary exploration.

The USB was an integrated system within the expanded Manned Space Flight Network (MSFN) and required the use of 30ft (9m) and 85ft (26m) antennae in California, Australia and Spain for complete coverage on a continuous basis, allowing MSFN stations to acquire a signal in sequence as the Earth rotated. Each spacecraft had a dedicated frequency pair so that two could be used at the same time. As with the Apollo CSM, the LM was assigned a frequency pair on a coherent ratio of 221/240 and could operate in either phase-modulation or frequency-modulation but not both. The CSM had two separate transmitters and so

The LM Ascent Propulsion System, with a fixed-thrust rocket motor designed to lift the Ascent Stage off the Descent Stage and back into lunar orbit. It was not used on Apollo 13. (Grumman)

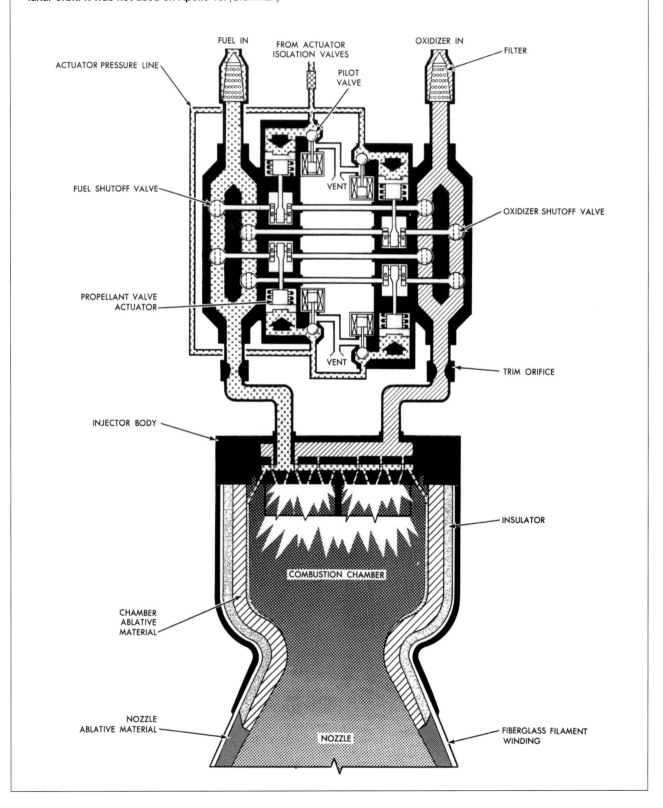

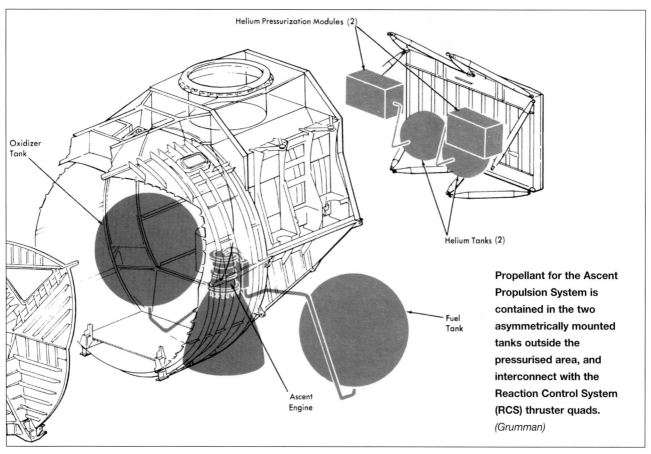

Helium Pressurization Modules (2)

Oxidizer
Tank

Helium Tanks (2)

Fuel
Tank

Ascent
Engine

Propellant for the Ascent Propulsion System is contained in the two asymmetrically mounted tanks outside the pressurised area, and interconnect with the Reaction Control System (RCS) thruster quads. *(Grumman)*

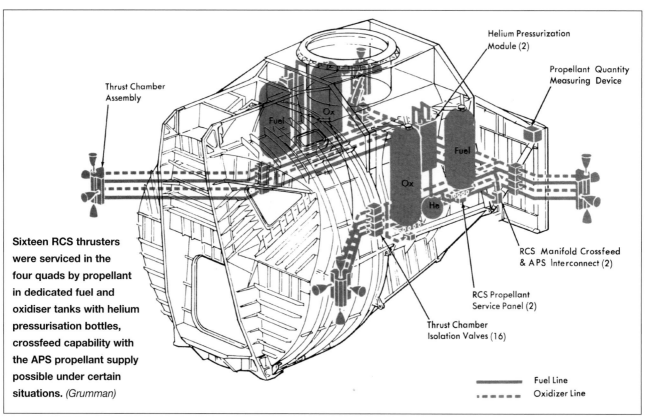

Helium Pressurization
Module (2)

Propellant Quantity
Measuring Device

Thrust Chamber
Assembly

Ox

Fuel

Fuel

Ox

He

RCS Manifold Crossfeed
& APS Interconnect (2)

RCS Propellant
Service Panel (2)

Thrust Chamber
Isolation Valves (16)

Sixteen RCS thrusters were serviced in the four quads by propellant in dedicated fuel and oxidiser tanks with helium pressurisation bottles, crossfeed capability with the APS propellant supply possible under certain situations. *(Grumman)*

———————— Fuel Line
- - - - - - - Oxidizer Line

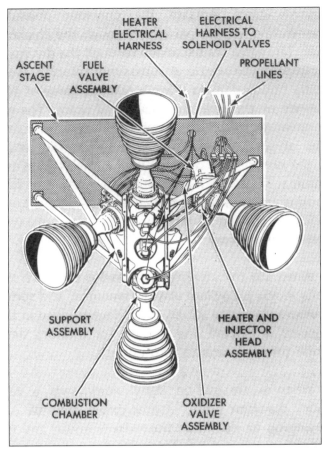

ASCENT STAGE

FUEL VALVE ASSEMBLY

HEATER ELECTRICAL HARNESS

ELECTRICAL HARNESS TO SOLENOID VALVES

PROPELLANT LINES

SUPPORT ASSEMBLY

HEATER AND INJECTOR HEAD ASSEMBLY

COMBUSTION CHAMBER

OXIDIZER VALVE ASSEMBLY

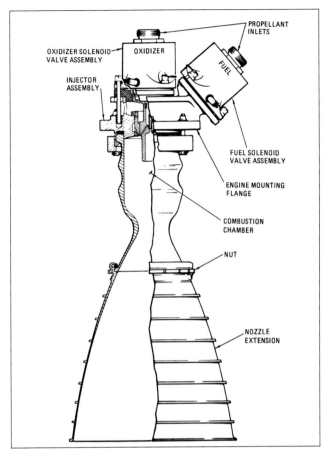

OXIDIZER SOLENOID VALVE ASSEMBLY

INJECTOR ASSEMBLY

OXIDIZER

PROPELLANT INLETS

FUEL

FUEL SOLENOID VALVE ASSEMBLY

ENGINE MOUNTING FLANGE

COMBUSTION CHAMBER

NUT

NOZZLE EXTENSION

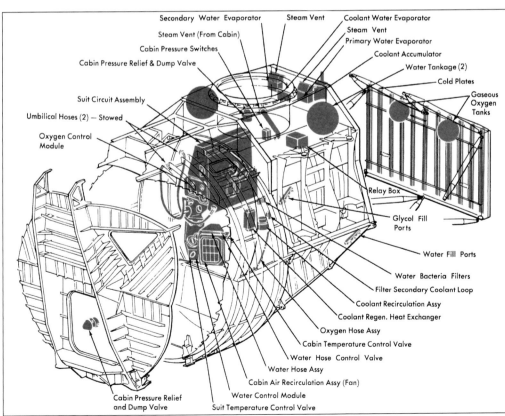

Secondary Water Evaporator
Steam Vent (From Cabin)
Cabin Pressure Switches
Cabin Pressure Relief & Dump Valve
Suit Circuit Assembly
Umbilical Hoses (2) — Stowed
Oxygen Control Module
Steam Vent
Coolant Water Evaporator
Steam Vent
Primary Water Evaporator
Coolant Accumulator
Water Tankage (2)
Cold Plates
Gaseous Oxygen Tanks
Relay Box
Glycol Fill Ports
Water Fill Ports
Water Bacteria Filters
Filter Secondary Coolant Loop
Coolant Recirculation Assy
Coolant Regen. Heat Exchanger
Oxygen Hose Assy
Cabin Temperature Control Valve
Water Hose Control Valve
Water Hose Assy
Cabin Air Recirculation Assy (Fan)
Water Control Module
Suit Temperature Control Valve
Cabin Pressure Relief and Dump Valve

ABOVE LEFT RCS thrusters are radiation-cooled motors used for attitude control and for translation manoeuvres. *(Grumman)*

ABOVE Cutaway of the principal design features of an RCS thruster. *(Grumman)*

LEFT The Environmental Control System (ECS) provided a clean and pressurised atmosphere of pure oxygen for the crew, and controlled temperatures for crew and equipment, the latter through a Heat Transport System (HTS). *(Grumman)*

could operate in both phase- and frequency-modulation at the same time.

LM-7 had an uplink frequency of 2101.802083MHz, and the downlink was fixed at 2282.5MHz. Phase-modulation was the norm for downlinks and has a constant amplitude irrespective of modulation, and thus allows the use of nonlinear RF amplifiers which are more efficient than those that must maintain linearity.

Two omnidirectional S-band antennae were carried on the forward and aft facing surfaces of the Ascent Stage. The one on the front was attached to the top of the faceplate; the other was attached to the aft equipment bay. A 26in (66cm) diameter steerable S-band antenna was attached to struts off the upper midsection on the +Y side of the LM. This incorporated a point source feed consisting of a pair of cross-sleeved dipoles over a ground plane. The antenna could be moved 174° in azimuth and 330° in elevation with a capture angle of +/-12.5° and once captured could track automatically.

Although not used on Apollo 13, LM-7 carried a deployable VHF antenna in the top of the Ascent Stage for communication with the astronauts on the surface during EVA (Extra-Vehicular Activity). Also stowed in the Descent Stage of LM-7 was an erectable S-band antenna. Comprising a 10ft (3.05m) diameter parabolic dish that opened in the form of an umbrella and was placed on the surface for relaying TV pictures direct to Earth, it too was not used on Apollo 13.

To outward appearance the Ascent Stage appeared flimsy, wrapped in thermal blankets to protect the structure and the equipment from temperature extremes and micrometeoroids. To protect against temperature extremes, the LM had both active and passive protection, the former in the Heat Transport System and its coolant loops discussed earlier. Passive protection was provided by combined thermal and micrometeoroid blankets which totally enclosed the vehicle. Fibreglass stand-offs

BELOW The Heat Transport System used a series of cold rails and cold plates, through which was passed a water-glycol solution to remove excess heat and discharge it to the vacuum of space through a sublimator, or provide heat through a transport loop to the systems and equipment on both Ascent Stage and Descent Stage. *(Grumman)*

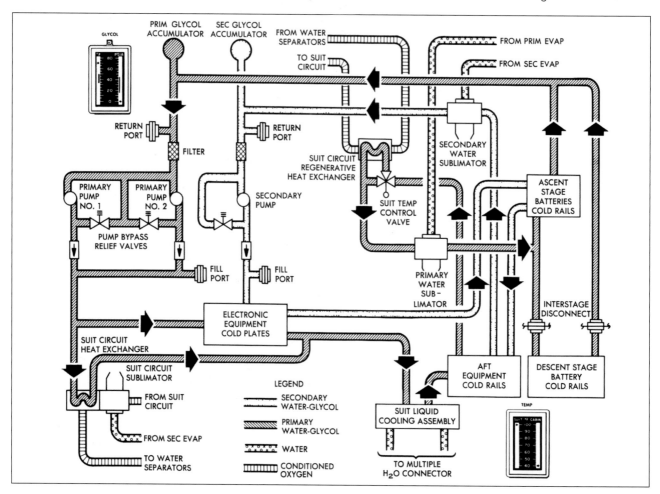

with low thermal conductivity held the blankets and isolated the main structure and externally-mounted elements.

Aluminium frames around the Ascent Stage propellant tanks prevented direct contact between the blankets and the structures. The thermal blankets comprised 25 layers of aluminised Mylar or H-sheet film, each layer being only 0.00015in (0.004mm) thick. The outer sheet was treated with a microscopically thin film of aluminium and the polymide sheets were hand-crinkled for maximum thermal venting. Where structures penetrated the shielding, struts were covered in the insulation with layers taped in H-film.

Externally, the Ascent Stage for LM-7 was 14.1ft (4.3m) in breadth across the extremities of the propellant tanks, 13.25ft (4.04m) in depth from front to back and had a height of 9.5ft (2.9m) attached to the top of the Descent Stage. It had an empty mass of 4,551lb (2,064kg) and carried 5,861lb (2,658kg) of propellant in the two APS tanks and the RCS system, 56% of the total mass of the Stage. In addition it had 2.67lb (1.21kg) of oxygen in the two Ascent Stage tanks and 84.2lb (38.19kg) of water in its two systems. With other sundry items, the total mass of the LM-7 Ascent Stage at launch was 10,502lb (4,764kg), approximately one-third of the total weight of *Aquarius*.

The Descent Stage

The general configuration of the Descent Stage comprised two pairs of parallel beams arranged in a cruciform with structural upper and lower decks. These formed five box-shaped enclosures in the form of a central box surrounded by four other boxes of equal volume each located at 90° to the other. When connected by struts between the ends of the main beams, four quadrants were formed. When viewed in plan these formed the shape of an irregular octagon. These quadrants were numbered in sequence, with No 1 between the +Z and -Y axes (left front), No 2 between the -Y and the -Z axes (left rear), No 3 between the -Z and +Y axes (right rear) and No 4 between the +Y and +Z axes (right front).

The primary structure consisted of aluminium alloy, chemically milled webs, and extruded and milled stiffeners with capstraps. The ends of the beams were closed off by bulkheads and the aluminium alloy tubular outrigger truss assemblies for the landing legs were attached at the ends of each pair of beams. The central box structure formed by the crossbeams housed the Lunar Module Descent Engine (LMDE), with compartments in the X and Y axis beams containing the oxidiser and fuel tanks respectively. The box structure had a depth of 5.75ft (1.75m) and a width of 13.8ft (4.21m). Two of four attachment fittings for the Ascent Stage were situated on the forward compartment beams, while the other two were on the aft beam of the side compartments.

In addition to two batteries, Quad 4 contained the Modularized Equipment Stowage Assembly (MESA), a drop-down tray of equipment used by the astronauts on the surface of the Moon. Diametrically opposite at the left rear of the Descent Stage, Quad 2 held a water tank and the Apollo Lunar Surface Experiments Package (ALSEP), an array of scientific instruments which Lovell and Haise were to have laid out across the lunar surface after they landed.

The exterior of this quadrant also carried a graphite cask within which was a radioisotope thermoelectric generator (RTG) for providing electrical power from the radioactive decay of plutonium 238 dioxide. For this reason the quadrant was referred to as the Scientific Equipment Bay. The plutonium pellet would have been extracted by an astronaut using a special tool and inserted into the RTG, which would be carried to the ALSEP deployment site to power the instruments.

Quad 3 at the right rear carried a retro-reflector pack of mirrors laid out in the form of a rectangular egg tray by which laser beams from Earth would be reflected back for precise distance measurements. It also held supplementary equipment including the gaseous oxygen tank, a helium pressurisation bottle, an ambient helium bottle and landing radar electronics. On later missions the Lunar Roving Vehicle would be installed, folded up, in Quad 1. For Apollo 13 it held the S-band erectable antenna for use on the lunar surface, two batteries and an electronic control assembly.

The cantilevered landing gear consisted of four assemblies each connected to an outrigger that extended from the ends of each parallel beam, with struts, trusses, a footpad and deployment mechanisms. A nine-rung ladder was attached to the forward (+Z) leg. The gear could survive a landing rate of 10ft/sec (6.8mph or 3.05m/sec/10.9kph), a 7ft/sec (4.8mph or 2.13m/sec/7.7kph) lateral velocity and a vertical attitude within 6° of vertical. With the exception of the forward landing leg, the other three legs had a contact sensing probe that extended below the footpad to about 5ft (1.5m) and advised the crew to shut down the descent engine.

The leg consisted of a primary strut attached to the outrigger at one end and supporting the footpad at the other. It was a piston-cylinder arrangement with a maximum 32in (81.3cm) compression stroke, attenuated by a one-time crushable honeycomb cartridge in each strut. The footpad consisted of a 37in (94cm) aluminium-honeycomb pan attached by a ball fitting.

At touchdown, with only minimal propellant reserves remaining in the Descent Stage, LM-7 would have had a mass on the surface of about 16,600lb (7,530kg). At a total surface area of 1,075in^2 (6,936cm^2) for all four pads, when converted at a lunar surface gravity of 5.3ft/sec^2 (1.62m/sec^2), about one-sixth that of the Earth, the four pads would have exerted a pressure of 2.65lb/in^2 (18.27kPa) at the surface. This intriguing fact is why LM footpads appeared to rest on the surface of a loosely compact regolith (lunar soil), depressing it by a relatively small amount and not digging in. Of course, LM-7 would never reach the surface.

The ladder fixed to the primary strut on the +Z leg was 20in (50.8cm) between rail centres, with nine rungs spaced 9in (23cm) apart. The top of the ladder led directly down from the platform, or 'porch', affixed to the outrigger assembly leading from the front hatch. The platform was 32in (81.3cm) wide and 45in (114.3cm) long, with a rail each side and a corrugated surface. The top rung on the ladder was 18in (45.7cm) below the front edge of the platform and the lowest rung lay about 30in (76.2cm) above the footpad.

Apart from providing a stable platform on the surface, the primary role of the Descent Stage was to provide propulsion for decelerating the LM from an orbital speed of about 3,800mph (6,114kph) to zero in less than 12 minutes,

descending from an altitude of about 10 miles (16.1km), during which it would travel about 185 miles (298km) across the lunar surface. Normally the DPS would consume around 96% of tanked propellant. LM-6, for instance, landed during Apollo 12 with 667lb (302.6kg) remaining, less than that for Apollo 11 (770lb/349kg), which had brought great tension as commentators suspected it was about to run dry.

The Lunar Module Descent Engine (LMDE) was advanced for its day. It achieved reliability through that most enduring of engineering ethics: simplicity and functionality over sophistication. It produced a nominal maximum thrust of 10,000lb (44,480kN) but in reality it can be operated at throttle levels between 9,870lb (43,902kN) and within a range of 1,050lb (4,670kN and 6,800lb (30,246kN) of thrust. Throttle ranges between 65% and 92.5% result in damage to the engine and are never used except in an emergency by manual override.

The engine itself weighed 360lb (163kg) and had a length of 85in (216cm) and a diameter of 59in (149.9cm) at the base of the engine skirt, which had an expansion ratio of 47.4:1 – that is, the area of the base of a circle prescribed by the radius of the skirt (2,734 in^2/17,640cm^2) is 47.4 times the area of the combustion chamber throat (57.68 in^2/372.15cm^2). The ratio of the exit area of the throat to the combustion chamber was 1:16. Nominal operating pressure was 103.4lb/in^2 (712.9kPa). For conservative use, due to material degradation the engine was limited to a burn duration of 910 seconds, more than enough for a landing.

The LMDE was mounted within the central box structure on a gimbal platform that allowed +/-6° of motion in the Z and Y axes to maintain the thrust line through the centre of gravity by modest amounts of pitch or roll. The bottom of the LMDE engine skirt was set 24in (61cm) below the underside of the Descent Stage, which would have brought the bottom of the skirt to within 20in (51cm) of the surface of the Moon, assuming a flat plane. In the event of a hard landing the skirt could crush up to 28in (71cm) without pushing up through the Descent Stage.

Propellants for the LMDE were in the Descent Propulsion System (DPS – pronounced 'dips'), and consisted of four tanks, one in each of the four box-shaped enclosures surrounding

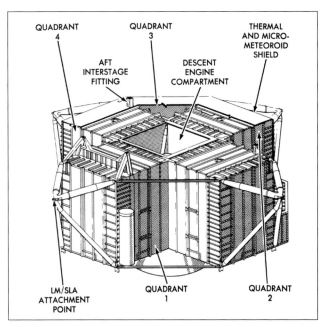

ABOVE One of two water tanks situated on opposing sides of the upper mid-section straddling the tunnel area, together with associated plumbing and wiring. *(Grumman)*

ABOVE RIGHT The structural purpose of the Descent Stage was to accommodate the Descent Propulsion System and its engine and propellant tankage, as well as affording a stable landing platform and launch pad for the Ascent Stage to return to the Apollo spacecraft at the end of a stay on the lunar surface. *(Grumman)*

BELOW The Descent Stage consisted of machined parts and chemically milled panel/stiffener assemblies mechanically fastened. Fabrication began with joining of the machined picture frames and the chemically milled assemblies forming the engine compartment. Outrigger bulkhead assemblies are attached to the engine compartment with machined cap strips, the eight remaining panel/stiffener assemblies, the upper and lower machined decks and the machined interstage fittings added to the structure. The Descent Stage also supported five major electrical harnesses and 45 cable assemblies. *(Grumman)*

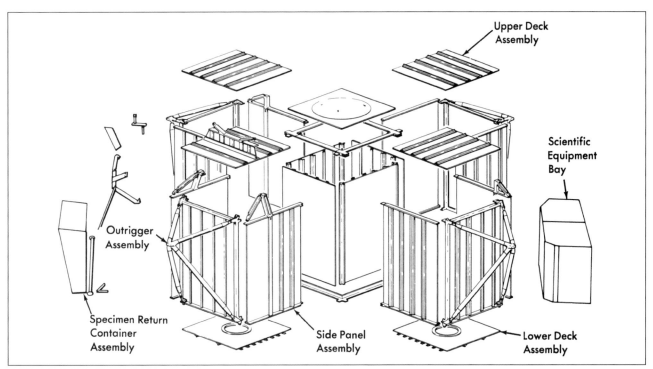

the central box where the engine was mounted. Each pressure vessel was approximately spherical, with a diameter of 51in (129.5cm) and length of 70.8in (179.8cm), fabricated in 6A1-4V titanium steel.

For LM-7 the two oxidiser tanks each held a nominal 5,645lb (2,560.6kg) of nitrogen tetroxide, while the two fuel tanks each held 3,524lb (1,598.5kg) of Aerozene-50. The horizontally opposed oxidiser tanks were located in the +Z and -Z bays, the two fuel tanks in the +Y and -Y bays, and identical tanks were fed in parallel with balance lines at the top and bottom of each tank.

Pressurisation of the propellant tanks was the responsibility of the cryogenic supercritical helium (SHe), about which much debate raged before and during the flight. While it promised a weight saving of 280lb (127kg) over a gaseous system, problems with using the SHe for an initial engine start pushed the need for an ambient helium supply to get the engine going. So a sphere, 14.9in (37.85cm) in diameter, was installed in Quad 3 in the general area of the SHe tank to hold 1.12lb (0.51kg) of helium in a pressure of 1,600lb/in² (11,032kPa) at 70°F (21.1°C).

Other consumables carried in the Descent Stage included a water tank and an oxygen tank. The water tank was installed in Quad 2. With a diameter of 28.48in (61.18cm) and a length of 32.5in (82.55cm), it was fabricated from 6061-T6 aluminium and contained 253.6lb (115kg) of water. The total amount of water aboard LM-7 was sufficient to recharge the Personal Life Support System (backpacks) of each astronaut with 9.15lb (4.15kg) of water.

The gaseous oxygen tank in the Descent Stage was installed in Quad 3 and consisted of a sphere 21.72in (55.17cm) in diameter, fabricated from D6AC steel, containing 49.3lb (22.36kg) of oxygen. It was stored at 2,730lb/in² (18,823kPa). Here too the quantity was determined by the Moonwalks, in this case the need to repressurise the Ascent Stage twice after each of the two EVAs, having twice vented the cabin atmosphere to a vacuum to start the Moonwalks. Again, the much greater quantity of oxygen than that required merely to maintain cabin pressure was to prove invaluable during the flight of Apollo 13.

Electrical power for the LM was provided by six batteries, four of them in the Descent Stage.

Two of these were located on a cold plate in Quad 1 and two in Quad 2. Each silver-zinc battery was 16.5in (41.9cm) by 10.15in (25.8cm) by 9.56in (24.28cm) in size and weighed 135lb (61kg), with 20 silver-zinc cells and a potassium hydroxide electrolyte. Nominal voltage was 30V DC at 400Ah. The four Descent Stage batteries provided 73% of total electrical power available from the six batteries on *Aquarius*. Electrical power would prove one of the most demanding challenges during the flight of Apollo 13.

One item of equipment not used on *Aquarius* was the landing radar, situated on the underside of the Descent Stage. It could transmit three Doppler velocity beams at 10.51GHz and an altitude radar beam at 9.58GHz. Four broadside arrays could receive the reflected energy and an electronics assembly would have processed the Doppler and continuous-wave FM returns for velocity and slant-range data. The radar would have been switched on at 50,000ft (15,240m) during descent and altitude data was expected by 25,000ft (7,620m), although experience had shown it to be effective at 40,000ft (12,192m) – none of which was of any use to Apollo 13.

For Apollo 13 the Descent Stage of LM-7 had a dry mass of 4,359lb (1,977kg) and carried 18,338lb (8,318kg) of propellant, 84% of Stage weight. The total launch weight of the Descent Stage was 23,001lb (10,433kg) including 7,048lb (3,197kg) of Aerozene-50 fuel and 11,290lb (5,121kg) of nitrogen tetroxide oxidiser. In addition it carried 49.3lb (22.36kg) of oxygen and 253.6lb (115.03kg) of water, the total mass contributing approximately two-thirds of the weight of the LM at launch. At lift-off for Apollo 13, LM-7 (*Aquarius*) had a mass of 33,503lb (15,197kg).

The LM was more capable than is often acknowledged. Much has been made in popular literature of how close the Apollo 11 crew came to running out of propellant. In fact, *Eagle* had 63.5 seconds of hover time remaining when it landed. Real-time readouts were in error by 36 seconds due to propellant sloshing in the tanks giving a false indication of quantity. Hence the myth that the crew had only 27 seconds of flight time remaining when they touched down. Anti-slosh baffles were developed and tested in time to get them aboard LM-8 for Apollo 14.

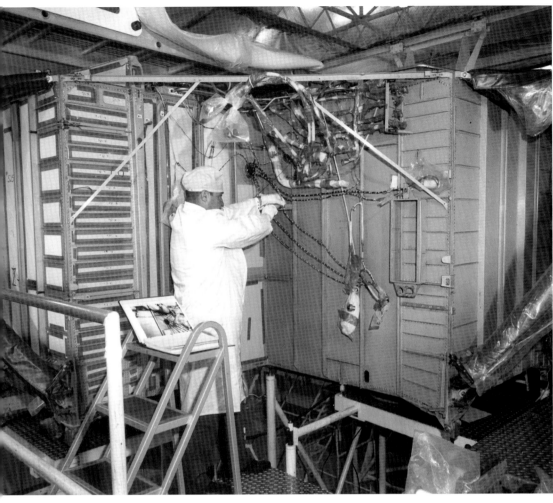

LEFT A technician works on wiring in one of the Descent Stage quads, with shear-web fittings for the lower landing-gear outrigger support assembly shown at lower right. *(Grumman)*

BELOW The configuration of the Descent Stage is built around the descent propulsion elements, including the Lunar Module Descent Engine (LMDE) and the propellant tanks, plus pressurisation equipment. Note the location of the supercritical helium (SHe) tank, which posed a frustrating annoyance before and during the flight. *(Grumman)*

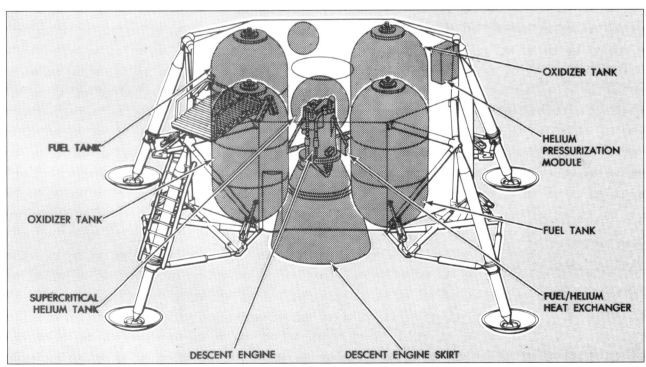

OXIDIZER TANK

HELIUM PRESSURIZATION MODULE

FUEL TANK

FUEL/HELIUM HEAT EXCHANGER

FUEL TANK

OXIDIZER TANK

SUPERCRITICAL HELIUM TANK

DESCENT ENGINE

DESCENT ENGINE SKIRT

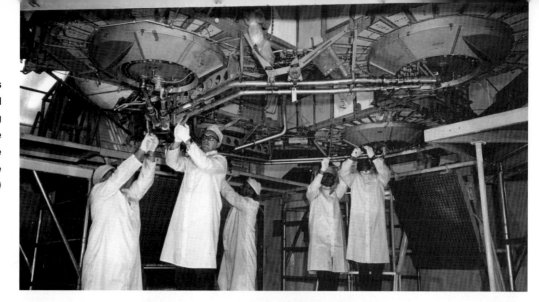

RIGHT Technicians install the pipes and conduits connecting the four Descent Stage propellant tanks. The engine has yet to be installed. *(Grumman)*

RIGHT The underside of the Descent Stage, showing the LMDE and the location of the landing radar as a rectangular plate. *(Grumman)*

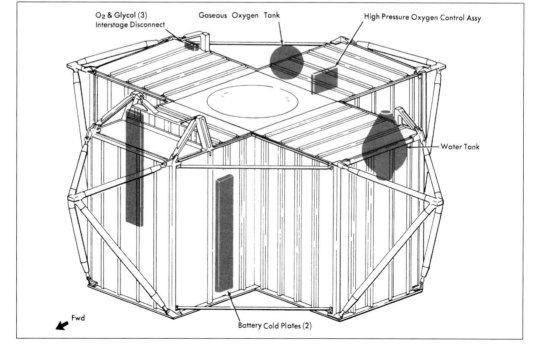

O₂ & Glycol (3) Interstage Disconnect

Gaseous Oxygen Tank

High Pressure Oxygen Control Assy

Water Tank

Fwd

Battery Cold Plates (2)

RIGHT Elements of the Environmental Control System located in the Descent Stage included a water tank, as well as oxygen systems and four batteries with associated cold plates. *(Grumman)*

ABOVE LEFT Not yet mated, but in the relative position on top of the Descent Stage, the Ascent Stage displays the forward-mounted rendezvous radar antenna, clearly showing cable runs across the top of the forward crew compartment. *(Grumman)*

ABOVE In this front view of the two stages, the No 4 quad displays the fittings for a cold plate for two of four Descent Stage batteries and its electrical control assemblies. *(Grumman)*

LEFT While the Aerozene-50 tank on the Ascent Stage focuses the eye in this picture. An interesting feature to note is the cold plate for one of the two Descent Stage batteries in the No 1 quadrant. Also, the EVA (Extra Vehicular Activity) rail attached to the left side of the front hatch on the forward face assembly, which would be used for astronauts to egress from the LM and work their way around to the side hatch on the Command Module in the event that the docking tunnel was blocked. *(Grumman)*

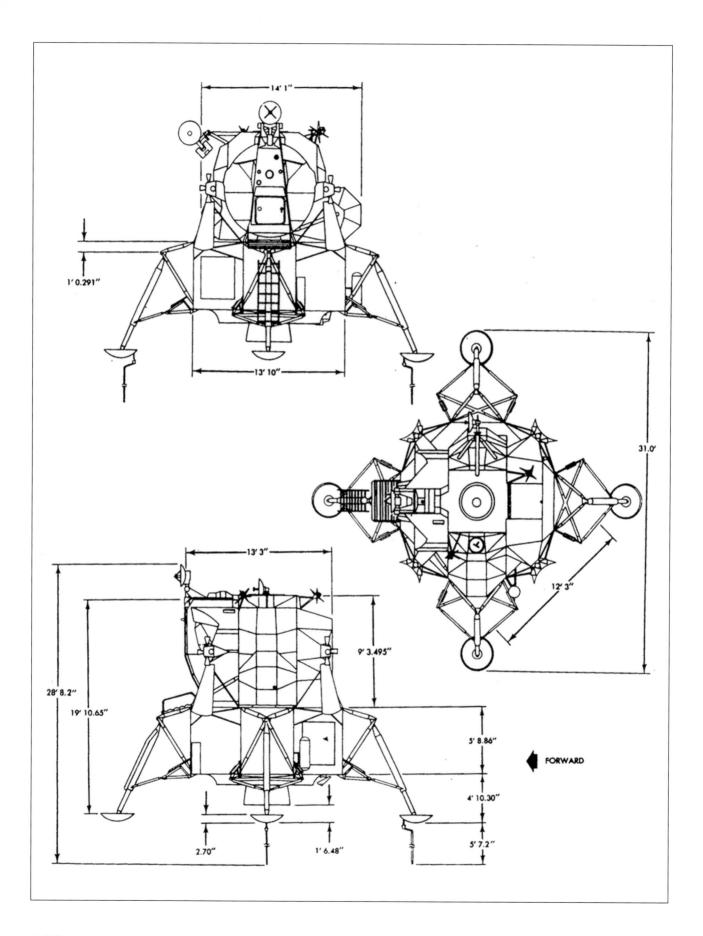

FORWARD

ABOVE The two mated stages of a Lunar Module, supported in a ground frame. Designed only to stand on the surface of the Moon in one-sixth gravity, it was incapable of supporting its own weight on Earth. *(Grumman)*

OPPOSITE General arrangement drawing of the LM with selected dimensions. *(Grumman)*

LEFT Astronaut Frank Borman explains the fundamentals of a Lunar Module to the UK Minister of Technology in the Wilson government, Tony Benn. *(NASA)*

Glossary

A Amps.

ACA Attitude Controller Assembly.

AEA Ascent Engine Assembly.

AGC Apollo Guidance Computer. Also Abort Guidance Computer.

AGS Abort Guidance System. Pronounced 'aggs'.

Ah Amp hours (battery capacity).

ALSEP Apollo Lunar Surface Experiments Package.

AMPTF Apollo Mission Planning Task Force.

AOS Acquisition of signal.

AOT Alignment Optical Telescope.

APS Ascent Propulsion System; also auxiliary propulsion system.

ARIA Apollo Range Instrumentation Aircraft.

AS-508 Apollo Saturn 508 (the eighth Saturn V to fly).

ASA Abort Sensor Assembly.

BPC Boost Protective Cover.

Btu/hr British thermal units per hour.

C&W Caution-and-warning sensor.

Capcom Capsule communicator. Astronaut at mission control who communicates with the spacecraft.

CDDT Count-Down Demonstration Test.

CDR Commander.

CECO Centre engine cut-off.

CM Command Module.

CMC Command Module Computer.

CMP Command module Pilot.

CO₂ Carbon dioxide.

COAS Crew Optical Alignment Sight.

CONTROL Control officer, responsible for guidance and navigation systems in the LM.

CSM Command Service Module.

DAP Digital Auto-Pilot.

DC Direct current.

DEDA Data Entry and Display Assembly. Pronounced 'deeda'.

ΔV Delta-v, *ie* change in velocity with direction.

DPS Descent Propulsion System. Pronounced 'dips'.

DSKY or disskee Display keyboard of the guidance and navigation computer.

ECS Environmental Control System.

EECOM Electrical, environmental and consumables manager.

EI Entry interface, and time to (+) or from (-) entry interface.

EMS Entry Monitor System.

Equigravisphere Point where the gravity fields of Earth and the Moon balance.

EST Eastern Standard Time.

EVA Extra-Vehicular Activity.

FAO Flight activities officer.

FDAI Flight Director Attitude Indicator.

FIDO Flight dynamics officer. Monitored and calculated flight path and trajectory.

FMEA Failure Mode & Effects Analysis.

g Measure of gravitational force.

G&NC Guidance and navigation computer.

GET Ground elapsed time in hours, minutes and seconds or just hours and minutes after launch.

GHz Gigahertz (billions of cycles per second).

GNC Guidance, navigation and control systems officer.

GUIDO Guidance officer. Monitored spacecraft computers and defined position of the spacecraft in space.

HGA High-Gain Antenna.

HTS Heat transport system.

IMU Inertial Measurement Unit.

INCO Instrumentation and communications systems officer.

kJ/hr Kilojoules per hour.

kN Kilonewtons force.

kPa Kilopascals.

kph Kilometres per hour.

KSC Kennedy Space Center.

L/D Lift over drag.

LEB Lower equipment bay.

LES Launch Escape System.

LGC Lunar Guidance Computer in the Lunar Module's PGNCS.

LiOH Lithium hydroxide.

LM Lunar Module.

LMDE Lunar Module Descent Engine.

LMP Lunar Module Pilot.

LOS Loss of signal.

LPD Landing Point Designator.

m Metres (distance).

Mach The speed of sound.

MCC Midcourse correction, MMC-1 being midcourse correction 1, MMC-2 midcourse correction 2, and so on.

MCC-H Mission Control Houston.

MER Mission Evaluation Room.

MESA Modularized Equipment Stowage Assembly.

MHz megahertz (millions of cycles per second).

MOCR Mission Operations Control Room. Pronounced 'moker'.

mph Miles per hour.

msec Milliseconds.

MSFN Manned Space Flight Network.

MSOB Manned Spacecraft Operations Building.

N Newtons force.

NAA North American Aviation.

NASA National Aeronautics & Space Administration.

OECO Outer engine cut-off.

P37 'Return to Earth' abort program of the guidance and navigation computer.

P40 Program that governs guidance, navigation and control systems during a thrust manoeuvre.

P52 An optical alignment using the Command Module sextant and telescope to update the computer on any drift in the guidance platform.

Pa Pascals (units of dynamic pressure).

PC Expressed as PC+1 or PC+2, meaning one or two hours after pericynthion.

Pericynthion The point at which a spacecraft launched from Earth is closest to the Moon.

PGNCS Primary Guidance, Navigation & Control System. Pronounced 'pings'.

PIPA Pulsed Integrating Pendulous Accelerometer.

PLSS Personal Life Support System backpack. Pronounced 'pliss'.

PTC Passive Thermal Control.

RCS Reaction Control System.

RECOVERY Recovery officer.

REFSMMAT Reference For Stable Member Matrix. A mathematical method of determining a spacecraft's celestial orientation and location.

RETRO Retrofire officer. Monitored propulsive de-orbit burns and trans-Earth injection burns out of Moon orbit.

RF Radio frequency.

RP-1 Kerosene fuel.

RTG Radioisotope thermoelectric generator.

S-band Frequencies in the 2–4GHz band.

S-IC First stage of Saturn V rocket.

S-II Second stage of Saturn V rocket.

S-IVB Third stage of Saturn V rocket.

SHe Supercritical helium.

SLA Spacecraft Lunar Module Adapter.

SM Service Module.

SPAN Spacecraft Analysis, co-located with Mission Control Houston.

SPS Service Propulsion System.

TD&E Transposition, Docking and Extraction.

TELMU Lunar Module telemetry, environmental and electrical systems officer.

TLI Trans-lunar injection.

TTCA Thrust/Translation Controller Assembly.

UHF Ultra High Frequency.

USB Unified S-band system.

V Volts.

VAB Vehicle Assembly Building.

VHF Very High Frequency.

W Watts of electrical energy.

X-axis Vertical (yaw).

Y-axis Lateral (pitch).

Z-axis Longitudinal (roll).

Index

Useful contacts

British Interplanetary Society
27/29 South Lambeth Road
London SW8 1SZ
Tel 020 7735 3160
www.bis-spaceflight.com
Signatory to the International Astronautical Federation, the BIS was formed in 1933 with Arthur C Clarke an early member. The Society is open to all and has three publications available on subscription, two of which deal with space history. Holds regular meetings with lectures and has a library open to members.

The Science Museum
Exhibition Rd, South Kensington SW7 2DD, UK
Tel 0870 870 4868
www.sciencemuseum.org.uk
Contains a full space gallery with many relevant exhibits including the Apollo 10 Command Module and artifacts from the space programs of Europe, the UK, the US, Russia and China.

National Air & Space Museum
6th & Independence Avenue SW
Washington DC 20560, USA
Tel 001 202 633 1000
www.nasm.si.edu
The world's largest aerospace museum with numerous galleries and exhibits covering the space program and the Shuttle specifically.

Space Center Houston
1601 NASA Parkway, Houston, Texas 77058, USA
Tel 001 281 244 2100
www.spacecenter.org
Contains one of the world's largest collections of space exhibits with tours of Apollo-related facilities right alongside the NASA Johnson Space Center.

US Space and Rocket Center
One Tranquility Base
Huntsville, Alabama 35805, USA
www.spacecamp.com/museum
Contains one of the largest displays of Shuttle-related hardware and exhibits anywhere and provides educational opportunities with IMAX theatre.

Kansas Cosmosphere and Space Center
1100 North Plum
Hutchinson
Kansas 67501-1418, USA
www.cosmo.org
Contains the Apollo 13 Command Module completely restored after a 12-year program.

Web sites:

www.hq.nasa.gov/alsj/frame.html
A NASA archive of highly detailed information on all Apollo missions edited by Eric Jones and Ken Glover.

www.nasa.gov
NASA's home page for a wide range of web pages dedicated to all aspects of space flight and with access to detailed Apollo information.

http://history.nasa.gov
The official NASA history site with a large volume of detailed information available for download.

Acknowledgements

I would like to thank the many people who have influenced the production of this book, including Jonathan Falconer, who first introduced me to Haynes, and Steve Rendle for guiding the project through its many stages. To James Robertson, a very special thanks for design and layout and for working beyond the call of duty to prepare a finished product to a very high standard. To Mark Hughes, sincere gratitude for commissioning the one book that leaves the author's desk with a very small piece of him in every copy, culmination of a 40-year-long desire to tell the story as it was and how it felt. Finally, my love and affectionate gratitude to Ann, whose idea it was that now was the time to do it. Thank you for your support and for the many selfless hours you endured while I was away from your side preparing this book.